Basic Math Skills

Teacher's Edition

Basic Math Skills

Teacher's Edition

Jeffrey Slater *and* **Rick Ponticelli**
North Shore Community College

IRWIN

Burr Ridge, Illinois
Boston, Massachussetts
Sydney, Australia

ISBN 0–256–14767–1 (Teacher's Edition)
Printed in the United States of America
1 2 3 4 5 6 7 8 9 WCB 0 7 6 5 4 3 2 1 0

To the Teacher

Basic Math Skills, designed with the mature student in mind, is an arithmetic text for students who need to learn or review the basic facts and procedures of arithmetic so they can be prepared for future courses in business and mathematics. The last chapter serves as an introduction to algebra and functions as a bridge between an arithmetic course and a first course in algebra. This text can be used for a self-paced course of study or for a more traditional lecture format course. The aim has been to produce a flexible text for the teacher and the student.

It has been written with the firm belief that everyone can learn math, provided dedication and a positive attitude are brought to the endeavor and nurtured throughout. To this end, the teacher is all important. The patient guidance, good humor, and active support of the teacher are the critical elements that ultimately lead to a student's success.

General Organization

The text contains enough material for a one-semester course in basic mathematics (15 weeks). Topics covered are: Chapter 1, Whole Numbers; Chapter 2, Fractions; Chapter 3, Decimals; Chapter 4, Percents; Chapter 5, Introduction to Algebra.

Each chapter is divided into bite-sized units. The instructor should feel free to skip or combine units as desired. If necessary, instruction may even begin halfway through a chapter. The instructor can use the chapter test as a diagnostic tool to determine where to begin.

Chapter Organization and Features

Each chapter has from 10 to 16 units. Most units begin with a brief presentation of the concept, followed by a set of rules for conducting the arithmetic procedure. The unit then presents worked-out examples, followed by a similar practice problem. Answers to *all* practice problems are provided at the end of the book. Finally, an exercise set is included for the student to solve. At the end of many of these exercise sets business application problems have been included to provide realistic applications of the concept being studied.

- Within each chapter appears a unit entitled "How to Dissect and Solve Word Problems." These include applications of the arithmetic procedures from previous units. A problem-solving chart helps the student to organize the information from the problem and to articulate the process needed for solving the problem.
- *Chapter tests* are included at the end of each chapter. Their content parallels the content of the chapter in both substance and format.
- *Charts, graphics, and color* are used at critical junctures to enhance understanding and to communicate information more effectively.

- A *glossary of terms and phrases* provides both definitions and examples of terminology.
- A *check figures section* gives answers to

  Every practice problem.

  Every fourth item from a drill-type exercise.

  Every business application.

  Every odd-numbered item from "How to Dissect and Solve Word Problems" units.

  Every item from the chapter tests.

  Entire exercises formatted as charts.

Every effort has been made to produce an error-free text. If any errors remain, it would be greatly appreciated if the corrections were sent to:

Professor Rick Ponticelli
Mathematics Department
North Shore Community College
1 Ferncroft Road
Danvers, MA 01923

or

Mr. Richard Hercher
Senior Sponsoring Editor
Richard D. Irwin, Inc.
1333 Burr Ridge Parkway
Burr Ridge, IL 60521

Supplements

The Electronic Calculator Guide provides instruction in the 10-key calculator. It also reviews the touch method, includes speed drills, helps students apply new skills to business math word problems, and can be used in tandem with many of the units in this text. A *Test Bank* that contains a battery of two forms of tests for the units of each chapter is available. Midterm and final exams are also included in the test bank.

Teaching Suggestions

This text, while attempting to create a strong foundation for future study in math, does not pretend to have the perfect explanations or sets of procedures for the many mathematical concepts presented. Furthermore, each student brings a unique background, temperament, and learning style to the teaching–learning process. The teaching suggestions that follow are presented as supplementary or alternative techniques that can be used by the teacher as he or she sees fit.

CHAPTER

Unit 1

Stress the need for the use of a hyphen when writing compound numbers.

Make certain students understand the term *compound number* by asking them to write its definition and to give an example.

When writing a number in numeral form, the instructor can highlight the period name and show how the period name essentially becomes the comma.

For example:

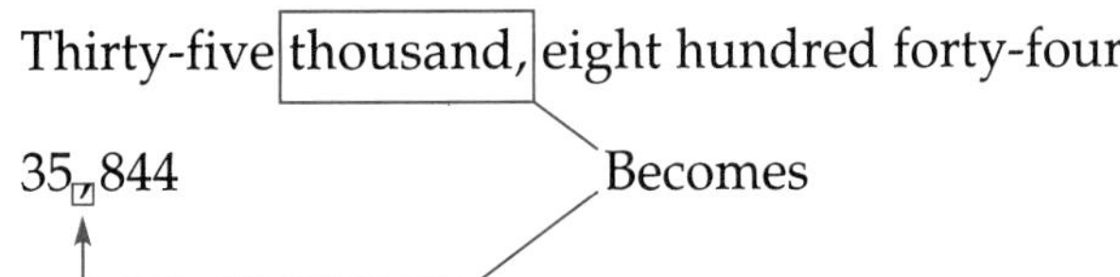

Unit 3

Explain to students that it is helpful to add a column of numbers by finding combinations of ten.

Unit 5

Tell students that rounding to the first digit is sometimes referred to as "rounding all the way."

Unit 7

Urge students to take a great deal of care with borrowing by writing down numbers neatly and showing each borrowing step explicitly.

Unit 9

Instead of using flash cards or a multiplication table to master basic multiplication facts, use a deck of cards (or two) with picture cards and aces removed. To drill multiplication, put two cards face-up on a table and state their product. Add a third card as needed.

Unit 10

Make sure students can identify partial products.

Unit 11

Explain how division by zero is impossible by relating division to multiplication. For example:

$\frac{4}{2} = 2$ because $2 \times 2 = 4$,

but

$\frac{4}{0} \quad 0$ because $0 \times 0 \quad 4$.

Discuss why division is *undefined* even though

$\frac{0}{0} = 0$ since $0 \times 0 = 0$.

Unless division by zero works under all circumstances we cannot call it a defined procedure.

CHAPTER 2

Unit 17

Make sure students know the terms *numerator, denominator, mixed numbers, proper fractions,* and *improper fractions* as these terms keep cropping up in future units.

One way to remember that the denominator is the bottom number is to remember that *d* stands for *down.*

Unit 18

Remind students that there is nothing wrong with improper fractions, as the label might indicate. In fact, it is sometimes better to express certain quantities as improper fractions.

Unit 19

It may be useful to talk about prime number at this point. Explain that the numerator and denominator are both expressed as the product of prime factors, and that reducing is the act of removing common prime factors from the numerator and denominator.

Show how reducing a fraction is nothing more than canceling off the number 1. For example:

$$\frac{15}{35} = \frac{5 \times 3}{5 \times 7} = \frac{\cancel{5}}{\cancel{5}} \times \frac{3}{7} = 1 \times \frac{3}{7} = \frac{3}{7}$$

Unit 22

An alternative method for finding the lowest common denominator is:

Example: $\frac{1}{4} + \frac{1}{8} + \frac{1}{9} + \frac{1}{12}$

1. Copy the denominators and arrange them in a separate row.

   4 8 9 12

2. Divide the denominators in step 1 by prime numbers. Start with the smallest prime number that divides into at least two of the denominators. Bring down any number not divisible.

$$\begin{array}{r|rrrr} 2 & 4 & 8 & 9 & 12 \\ \hline & 2 & 4 & 9 & 6 \end{array}$$

   The 9 was brought down because it was not divisible by 2.

3. Continue this process until no prime number will divide evenly into at least two numbers.

2	4	8	9	12
2	2	4	9	6
3	1	2	9	3
	1	2	3	1

4. To find the LCD, multiply all the numbers in the divisors and in the last row.

$$\boxed{2 \times 2 \times 3} \times \boxed{1 \times 2 \times 3 \times 1} = 72$$

Divisors   Last Row

Unit 25

An alternative method for borrowing with fractions is:

Example: $3\frac{1}{4} - 1\frac{2}{3}$

1. Find the LCD and raise numerators.

$$3\frac{1}{4} = 3\frac{3}{12}$$
$$-1\frac{2}{3} = -1\frac{8}{12}$$

2. Subtract 1 from the whole number of the top fraction and equate it to a fraction that has the same LCD as found in Step 1. Add this fraction to the top fraction.

$$3\frac{3}{12} = 2\frac{3}{12} + \frac{12}{12} = 2\frac{15}{12}$$
$$-1\frac{8}{12} = -1\frac{8}{12} = -1\frac{8}{12}$$

3. Subtract the fractions.

$$\begin{array}{r} 2\frac{15}{12} \\ -\ 1\frac{8}{12} \\ \hline 1\frac{7}{12} \end{array}$$

Units 27–30

Urge students to rewrite the reduced fractions that come about from multiple cancellations. Emphasize that working neatly will help to prevent errors.

CHAPTER 3

Unit 32

Tell students that the phrase *decimal fraction* and the word *decimal* mean the same thing.

Units 40 and 41

Units 40 and 41 can be presented shortly after Unit 27 as a way of emphasizing the relationship of fractions to decimals and vice versa.

Unit 42

Make sure students understand the idea of "outstanding deposits" and "outstanding withdrawals." One way to do this is to indicate that something out standing is something not yet received by the bank, and is thus out side the bank.

Unit 43

Students will wonder why employees are not being paid time and a half for working over 40 hours. Tell students that the rate for overtime varies from place to place, and for ease of computation, overtime has been disregarded; good discussion should ensue about whether or not this is realistic. Use this opportunity to let students experiment with different overtime rates.

CHAPTER

Unit 44

Emphasize that until the student actually "sees the hand move the decimal point two places," the conversion has not taken place.

Reinforce that changing decimals to percents and vice versa are shifts to the right or left by demonstrating the movement from *D* (for decimals) to *P* (for percent) and *P* to *D* in the alphabet.

A . . . . . *D* . . → . . *P* . . . . . Z

Changing a decimal to a percent.

A . . . . . *D* . . ← . . *P* . . . . . Z

Changing a percent to a decimal.

Unit 46

When determining which number is the base quantity, tell students to be mindful of the problem's wording. The base quantity is the number *immediately* after the word *of*.

Unit 48

Urge students to think about percent of increase and decrease in original base quantities. For example: If a $50,000 salary is increased by 3%, then the final salary is 103% of the original.

CHAPTER 5

Unit 55

Use the number line early and often as a way of demonstrating adding and subtracting signed numbers.

Subtracting signed numbers becomes terribly confusing for some students. De-emphasize the use of the word *subtraction.* Instead, consistently refer to it as "adding the opposite of."

Show the importance of parentheses as a means to making expressions visually less confusing by having students compare the expressions:

$$-7 + -3 - -5 + 16 - -12$$

and

$$-7 + (-3) - (-5) + 16 - (-12).$$

Ask which appears to be more effectively written and why. Ask students to insert parentheses where they feel they are necessary in an expression such as

$$5 - -2 + -22 + -14 - -\frac{1}{7}.$$

Units 56 and 57

Make sure students do not confuse rules for multiplying and dividing with rules for adding and subtracting signed numbers. An excellent assignment, group activity, or quiz item (or all of the above) would be to ask students to clearly, concisely, and correctly write the rules for adding, subtracting, multiplying, and dividing signed numbers.

Unit 58

Reinforce the order of operations with the mnemonic device PEMDAS ⇒ Please Excuse My Dear Aunt Sally. Be careful, however. Some older students sometimes find this device childish and insulting. Furthermore, make sure students understand that multiplication and division are performed in whatever order they occur when reading the expression from left to right. This procedure is also true for adding and subtracting.

Order of operations is an excellent topic to demonstrate on a calculator. The possibility exists that one of the students in the class will own a calculator that does not perform the order of operations implicitly. This situation will surely generate a discussion about the use of a calculator. Emphasize that the calculator is simply a tool for speed and accuracy.

Unit 59

Work with students to develop a glossary of terms and their mathematical equivalents. For example:

"is decreased by" is symbolized by $-$

Unit 61

Indicate that distribution can be done from either side of the set of parentheses.

$$2(7 + 5) = 14 + 10 = 24$$

also

$$(7 + 5)2 = 14 + 10 = 24$$

General Study Suggestions

The suggestions that follow are compiled with the thought in mind that, over the entire course of study, many of these points of view can be discussed and reinforced. Learning mathematics, like learning any new language, is a formidable task for many students. Discussing these aspects of study will prove helpful to many students because it gives them a clear focus on how to go about successfully engaging, learning, and retaining the material. Discuss the importance of:

- Maintaining a positive attitude.
- Attending each class.
- Active participation and attention in every class.
- Doing homework every day and distributing the time one spends on math homework over an entire week.
- Repeat completion of exercises already completed. Drill and repetition assist learning and retention.
- Getting to know someone in class and working together in study groups.
- Listening first and then taking notes during a lecture.
- *Reviewing* and *rewriting* one's notes.
- Reading the text, noting questions on paper or in the margins, and asking for clarification at the next class.
- Scheduling one's time effectively and using it to study math.
- Asking questions when something is not understood. Be sure to emphasize that these questions can be asked before, during, and after class, if necessary.
- Seeking extra help when needed.
- Preparing for tests in a comprehensive fashion over the long term.
- Coming to class prepared, having previewed the section(s) to be covered in class on that day.
- Reading directions and problems on tests carefully.
- Working neatly.
- Being aware of allotted time on tests.
- Checking one's work.

Gather the facts	What am I solving for?	What must I need to know or calculate before solving the problem?	[illegible]
[illegible]	Total [illegible]	[illegible]	[illegible]

Basic Math Skills

Basic Math Skills

Jeffrey Slater *and* **Rick Ponticelli**
North Shore Community College

IRWIN

Burr Ridge, Illinois
Boston, Massachussetts
Sydney, Australia

Cover illustration by Michael Mazzier
Cover designer: Jeanne M. Rivera / Laura Gunther

Senior sponsoring editor: Richard T. Hercher, Jr.
Marketing manager: Janelle Calon
Developmental editor: Carol L. Rose
Project editor: Denise Santor
Production manager: Bob Lange
Interior designer: Jeanne M. Rivera / David Corona Design
Composition and art: David Corona Design
Typeface: 11/13 Palatino
Printer: Wm. C. Brown Communications, Inc., Manufacturing Division

Library of Congress Cataloging-in-Publication Data

Slater, Jeffrey, 1947-
Basic math skills / Jeffrey Slater and Rick Ponticelli. — Student ed.
p. cm.
Includes index.
ISBN 0-256-14766-3
1. Mathematics. I. Ponticelli, Rick. II. title.
QA107.S538 1994
513'.12—dc20 93-34090

Printed in the United States of America
1 2 3 4 5 6 7 8 9 0 WCB 0 9 8 7 6 5 4 3

Thanks to Shelley,

Love,

Jeff

To Diane,

Love,

Rick

To the Student

This text has been designed to assist you in learning or reviewing the basic facts and procedures of arithmetic, so you can possess the mathematical skills needed for future business courses and everyday life.

Naturally, the foundation for your success in this endeavor is your own drive and dedication. These qualities, coupled with hard work, effective use of the text, and the support of others, will surely lead to your success.

General Organization

The text contains material for a one-semester course in basic mathematics (15 weeks). Topics covered are: Chapter 1, Whole Numbers; Chapter 2, Fractions; Chapter 3, Decimals; Chapter 4, Percents; and Chapter 5, Introduction to Algebra.

Chapter Organization and Features

Each chapter has from 10 to 16 units. Most units begin with a brief presentation of the concept, followed by a set of rules for conducting the arithmetic procedure. The unit then presents worked-out examples, followed by a similar practice problem. Answers to *all* practice problems are provided at the end of the book. Finally, an exercise set is included for you to work out. At the end of many of these exercise sets, business applications problems have been included to provide realistic applications of the concept being studied.

- Within each chapter appear units entitled "How to Dissect and Solve Word Problems." These include applications of the arithmetic procedures from previous units. A problem-solving chart helps you to organize the information from the problem and to articulate the process needed for solving the problem.
- *Chapter tests* are included at the end of each chapter. Their content parallels the content of the chapter in both substance and format.
- *Charts, graphics, and color* are used at critical junctures to enhance your understanding and to communicate information more effectively.
- A *glossary of terms and phrases* provides both definitions and examples of terminology.
- A *check figures section* gives answers to

  Every practice problem.

  Every fourth item from a drill-type exercise.

  Every business application.

  Every odd-numbered item from "How to Dissect and Solve Word Problems" units.

Every item from the chapter tests.

Entire exercises formatted as charts.

How to Use the Text

The following six-step method for using the text is recommended:

1. Preview the topic to be studied by scanning the entire unit.
2. Read the mathematical explanation carefully.
3. Work through the example, comparing it with the set of rules presented earlier.
4. Work through the practice problem, referring to the set of rules and the example with which it is matched.
5. Reread and review the mathematical explanation and examples.
6. Work through the problems in the exercise set.

Also, redo entire sets or portions of exercise sets as necessary to aid your retention of concepts and skills.

Study Tips

Learning math is like learning a new language, and often it is a difficult task. To minimize your frustration, it is important to maintain some fundamental study habits. Having good study habits will help you to stay focused on the task of learning this new material and will, in the long run, lead to your overall success.

The following list of suggestions is, by no means, a comprehensive one, but it does contain many key ingredients that lead to effective study.

- Find a quiet place to study where you will be exposed to a minimum of distractions. Try as often as possible to study in a library or in a math lab where assistance may be available to you. Be sure to equip your study area with all appropriate materials. If you study away from home, keep a list of items you will need to bring along with you.
- Try to work at a time of day or evening when you are at a peak energy level.
- Manage your time so that you study the material for at least 60–90 minutes each day. Make a weekly study schedule with each study session specifically identified. Then, stick to it! DON'T PROCRASTINATE!
- Attend every class. If an absence is avoidable, confer with your teacher to make sure you understand what material you are responsible for.
- In class, stay focused on the teacher and the content of the lecture, discussion, or activity. Actively involve yourself by thinking, asking questions, and offering answers when asked.
- Listen first and then take notes. Try to develop some speed writing skills to help your overall note-taking abilities. Avoid taking notes on material that is in the text but which the teacher is choosing to write on the blackboard anyway. Simply write any additional information in the margin of your text.

- Rewrite and review your notes after class as soon as possible. This procedure aids your retention of the content, and "neatens" up your notes, making them easier to follow.
- Make friends with other students in class as soon as possible. Study with these individuals. Mutual support and encouragement goes a long way in helping one to be a successful student.
- Politely ask questions in class when necessary. If the pace is too fast, request that the procedure be repeated. If you still don't understand after a repeat clarification, see the instructor after class for further help. Never feel that you are a burden for asking questions.
- Drill and repetition are important in mathematics. Repeatedly complete exercises even if they have already been done. This habit ultimately leads to increased skill.
- Read the text actively. Ask questions of what you read. Reflect or ponder about what you have read. If you have questions about the content of the text, ask your instructor. Try to preview the next class lecture by reading the upcoming section before class.
- Be sure to understand the vocabulary of mathematics. If you only think you understand a word or phrase, be certain to look it up. The language of math is very precise. Don't overlook this aspect of your study.
- Prepare for a test well in advance of the test itself by making a test preparation schedule for yourself.
- To prepare for tests, review notes, previous tests or quizzes, textbook information, and handouts the instructor provided.
- Be sure that you understand what a test will cover, how much time will be allotted for its completion, and what its format will be.
- Work swiftly on the test. Be sure you understand the questions and directions. Ask for clarification when necessary. Always be aware of the time, and do not get bogged down in unproductive problem solving or aimless procedures.
- If time allows, always check your work. Never change an answer because of a sense of anxiety. Only change an answer if you are *certain* you must!
- Turn any negative thoughts into positive ones. For example:

  "Jeez, we never did one like this; what the heck is this doing on the test? Gimme a break, willya!"

  instead,

  "Jeez, this doesn't look familiar. That's OK, though. I know this material, so I'll be able to figure it out. Let me take a closer look at this."
- Most important, try to keep a positive and receptive attitude about this learning experience.

Best of luck to you!

Acknowledgments

I would like to thank the many people who assisted in this project. In particular those who reviewed all or part of the manuscript by making many helpful suggestions and comments: Walter Reece, Four-C College; John Bloomquist, Lamson Junior College; Sandra Richards, Watterson College; Howard Frank, Palmer Business Institute; Penny O'Donnell, Pennsylvania Business Institute; and Rosemary Pisercio, College of San Mateo. I would also like to thank Professor John Tobey of North Shore Community College for his many helpful comments and Patricia Lovie for her professionalism and excellent word-processing skills. Special thanks are extended to Richard Hercher for his support and guidance, Carol Rose for her knowledgeable assistance and sense of humor, Denise Santor for her careful attention to this project, and Jeanne Rivera for her supervision of the design.

To my friends, my family, and anyone who ever expressed interest in this project. Finally, and most important, to Diane, Michael, and Mark; your love, encouragement, and patience throughout this project are deeply appreciated.

Rick Ponticelli

Contents

CHAPTER 4 Percents 131

CHAPTER 5 Introduction to Algebra 175

Basic Math Skills

CHAPTER 1

Whole Numbers

UNITS

UNIT

1 Writing Whole Numbers in Verbal and Numeral Form

A whole number written in numeral form uses place values. Working from right to left in a number, these place values are units, tens, hundreds, thousands, hundred thousands, millions, and so on. Each number is also separated into groups of three digits each, called **periods.** The diagram shows these place values and the period names.

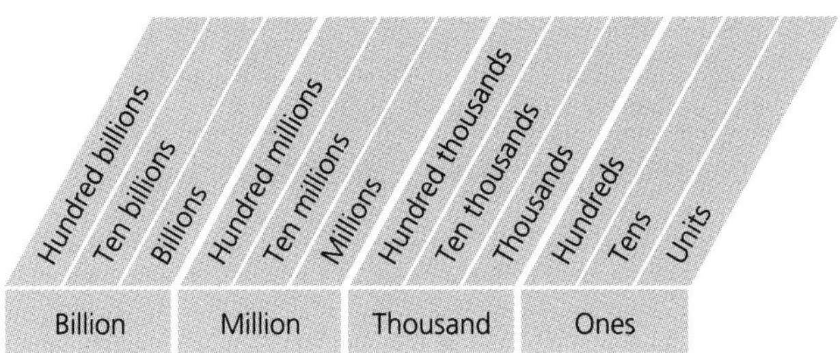

1 When reading a whole number, work from left to right, and simply state the number in each period, along with the name of the period. It is unnecessary, however, to give the name of the ones period when reading a number from it.

2 When writing a whole number in verbal form, use commas to separate each period. Use a hyphen when expressing the compound numbers 21–99. Thus, the number 7,585,376,491 would be written as: "seven billion, five hundred eighty-five million, three hundred seventy-six thousand, four hundred ninety-one."

3 To write a whole number in numeral form, write the number named, replacing each period name with a comma. Thus, the number eight million, five hundred seventy-six thousand, one hundred sixty-eight would be written as "8,576,168."

EXAMPLE 1 **Write 415,783 in words.**

Answer: Four hundred fifteen thousand, seven hundred eighty-three.

PRACTICE 1 **Write 783,696 in words.**

Answer: seven hundred eighty-three thousand, six hundred ninety-six

---

EXAMPLE 2 **Write six million, three hundred seventy-five thousand, eight hundred eighty-four in numeral form.**

Answer: 6,375,884

PRACTICE 2 **Write three million, one hundred ninety-five thousand, two hundred two in numeral form.**

Answer: 3,195,202

UNIT 1 Name ______________________ Date ______________

Writing Whole Numbers in Verbal and Numeral Form

Write the following numbers in words:

1. 37 — thirty-seven
2. 593 — five hundred ninety-three
3. 1,444 — one thousand, four hundred forty-four
4. 1,073 — one thousand, seventy-three
5. 486,992 — four hundred eighty-six thousand, nine hundred ninety-two
6. 406,371 — four hundred six thousand, three hundred seventy-one
7. 5,696,404 — five million, six hundred ninety-six thousand, four hundred four
8. 2,300,581 — two million, three hundred thousand, five hundred eighty-one
9. 27,686,014 — twenty-seven million, six hundred eighty six thousand, fourteen
10. 33,418,002 — Thirty-three million, four-hundred eighteen thousand, two
11. 402,989,115 — Four hundred two million, nine hundred eighty-nine thousand, one hundred fifteen
12. 16,147,317,296 — Sixteen billion, one hundred forty-seven million, three hundred seventeen thousand, two hundred ninety-six
13. 419,419,000,419 — Four hundred nineteen billion, four hundred nineteen million, four hundred nineteen

Write the following in numeral form:

14. Eight hundred forty-seven — 847
15. Fifteen thousand, six hundred eighteen — 15, 618
16. Sixty-two million — 62,000,000
17. Four hundred thirty-nine million, two hundred twenty-four — 439,000,224
18. Seven million, ninety-three thousand, five hundred two — 7,093,502
19. Sixteen million, seventy thousand, four hundred — 16,070,400
20. Thirty-nine billion, four hundred thirty-five million — 39,435,000,000
21. Fifty-five billion, eight hundred twenty-two million, six hundred six thousand, four hundred twenty — 55,822,606,420
22. Nine billion, three thousand, six — 9,000,003,006
23. One hundred ninety-four million — 194,000,000
24. Five hundred ten million, eight hundred forty-six thousand, one hundred eighty-two — 510,846,182
25. Seven million, six hundred ninety-three thousand, two hundred four — 7,693,204
26. Three billion, four million, fourteen thousand, four hundred fifty-six — 3,004,014,456

Business Applications:

27. The total volume of New York Exchange Bonds traded was 17,420,000. Write this number in words.

    Seventeen million, four hundred twenty thousand

28. The total stock shares traded on the New York Exchange was listed as 46,620,327. Write this number in words.

    Forty-six million, six hundred twenty thousand, three hundred twenty-seven

29. A check is written for the amount of fourteen thousand, seven hundred thirty-two dollars. Write this amount in numeral form.

    $14,732

30. A paycheck is forwarded to an employee in the amount of three thousand, six hundred forty-three dollars. Write this amount in numeral form.

    $3,643

UNIT

2 Rounding Whole Numbers

The process of rounding whole numbers is an effective tool for approximating arithmetic computations.

Whole numbers can be rounded to an identified digit within a number, or the identified digit can be the first digit of the number. The steps for rounding whole numbers are:

1 Identify the digit you want to round.

2 If the digit to the right of the identified digit is 5 or more, increase the identified digit by one and change all digits to the right of the identified digit to zeros.

3 If the digit to the right of the identified digit is less than 5, change that digit and all digits to its right to zeros. Do not change the identified digit.

EXAMPLE 1 **Round 9,367 to the nearest hundred.**

Answer:

1. The digit 3 is to be rounded.
2. The digit 6 is to its right. Because it is 5 or more, the identified digit 3 is rounded to 4.
3. The number 9,367 rounded to the nearest hundred is 9,400.

PRACTICE 1 **Round 7,471 to the nearest hundred.**

Answer: *7,500*

EXAMPLE 2 **Round 10,537 to the nearest thousand.**

Answer:

1. The digit 0 is to be rounded.
2. The digit 5 is to its right. Because it is equal to 5, the identified digit 0 is rounded to 1.
3. The number 10,537 rounded to the nearest thousand is 11,000.

PRACTICE 2 **Round 40,561 to the nearest thousand.**

Answer: *41,000*

EXAMPLE 3: **Round 83,176 to the nearest thousand.**

Answer:

1. The digit 3 is the identified digit.
2. The digit 1 is the digit to its right. Because it is less than 5, it is changed to zero, along with all the digits to its right.
3. The number 83,176 rounded to the nearest thousand is 83,000.

PRACTICE 3: **Round 16,409 to the nearest thousand.**

Answer: 16,000

UNIT 2 Name ______________________ Date ______________

Rounding Whole Numbers

Round the following numbers to the given value:

1.	87	ten	90
2.	217	ten	220
3.	4,517	hundred	4,500
4.	63,017	thousand	63,000
5.	63,710	thousand	64,000
6.	989,467	ten thousand	990,000
7.	451,638	ten thousand	450,000
8.	28,459	hundred	28,500
9.	14,178	hundred	14,200
10.	9,602,416	million	10,000,000
11.	3,726,000	thousand	3,726,000
12.	4,699,999	hundred thousand	4,700,000
13.	487,369	hundred	487,400
14.	56,487	ten thousand	60,000
15.	63,598	ten	63,600
16.	14,017	thousand	14,000
17.	93,707	ten	93,710
18.	10,119	thousand	10,000
19.	47,125,931	million	47,000,000
20.	520,404	thousand	520,000
21.	700,776	thousand	701,000
22.	19,167,700	ten thousand	19,170,000
23.	80,600	ten thousand	80,000

Business Applications:

24. An ad agency has sales of $749,143. Round this figure to the nearest ten thousand.

    $750,000

25. Over four hours, 3,198 shares of stock were traded. Round the number of shares traded to the nearest hundred.

    3,200 shares

UNIT

3 Addition of Whole Numbers

Numbers to be added in a group of numbers are called **addends.** The result of the addition is called the **sum** or **total.**

To add a column of numbers:

1. Add the numbers in the column from top to bottom.
2. Check the answer by adding the numbers from bottom to top.

The exercise that follows the examples below is intended to give practice with basic addition facts. Because effective addition of larger numbers depends on mastery of these facts, practice the exercise until you are confident you can swiftly and correctly complete it.

EXAMPLE 1 **Add: 3 + 6.**

$$\begin{array}{r} 3 \\ +\ 6 \\ \hline 9 \end{array} \qquad \text{Check: } \begin{array}{r} 3 \\ +\ 6 \\ \hline 9 \end{array}$$

Answer: 9

PRACTICE 1 **Add: 4 + 5.**

Check:

Answer: 9

EXAMPLE 2 **Add: 2 + 5 + 1.**

$$\begin{array}{r} 2 \\ 5 \\ +\ 1 \\ \hline 8 \end{array} \qquad \text{Check: } \begin{array}{r} 2 \\ 5 \\ +\ 1 \\ \hline 8 \end{array}$$

Answer: 8

PRACTICE 2 **Add: 3 + 7 + 2.**

Check:

Answer: 12

UNIT 3 Name ______________________ Date ____________

Addition of Whole Numbers

Add:

1.	4 + 4 8	5 + 8 13	7 + 4 11	8 + 9 17	9 + 2 11	1 + 1 2	9 + 5 14
2.	2 + 8 10	3 + 0 3	4 + 3 7	7 + 5 12	4 + 2 6	6 + 3 9	5 + 2 7
3.	3 + 1 4	2 + 1 3	5 + 3 8	6 + 9 15	9 + 5 14	8 + 3 11	5 + 1 6
4.	6 + 1 7	3 + 9 12	7 + 5 12	5 + 5 10	2 + 1 3	8 + 6 14	5 + 4 9
5.	9 + 8 17	2 + 2 4	8 + 8 16	9 + 4 13	9 + 9 18	8 + 7 15	7 + 2 9
6.	6 6 + 5 17	3 7 + 1 11	9 6 + 2 17	4 7 + 9 20	2 3 + 0 5	9 6 + 2 17	4 7 + 4 15
7.	4 6 + 9 19	5 4 + 8 17	2 1 + 3 6	6 5 + 1 12	8 3 + 5 16	2 8 + 5 15	6 2 + 9 17
8.	6 5 1 + 4 16	2 8 3 + 7 20	4 9 6 + 4 23	3 9 4 + 1 17	7 1 6 + 2 16	2 5 8 + 3 18	5 2 6 + 8 21

Business Applications:

9. Add the following hours worked by a part-time employee of the Lenox Company: Monday, 4; Tuesday, 6; Thursday, 3; and Saturday, 5.

   18 hours

10. Add the following amounts paid for tools at a local hardware: $4, $3, $5, and $2.

   $14

UNIT 4 Horizontal and Vertical Addition of Larger Numbers

To add a group of larger numbers:

1. Write the numbers to be added in column fashion, so units are vertically lined up with units, tens are lined up with tens, hundreds with hundreds, and so on.
2. Begin at the right and add the numbers in each column separately, writing the sum underneath each column.
3. If the sum of a column is greater than 9, write the right-hand figure of the sum underneath the column, and add the left figure to the next column of numbers.
4. Write the entire sum of the last column of numbers.

EXAMPLE 1 **Add: 309 + 43 + 824.**

$$\begin{array}{r} {}^{1}\\ 309 \\ 43 \\ +\ 824 \\ \hline 1{,}176 \end{array}$$

Answer: 1,176

9 + 3 + 4 = 16
Write 6 and carry 1 to the next column.
4 + 2 + 1 = 7.
Write 7.
8 + 3 = 11.
Write 11.

PRACTICE 1 **Add: 347 + 35 + 922.**

$$\begin{array}{r} 347 \\ 35 \\ +\ 922 \\ \hline \end{array}$$

Answer: 1,304

EXAMPLE 2 **Add: 205 + 834 + 286.**

Rewrite and add.

$$\begin{array}{r} {}^{1\,1}\\ 205 \\ 834 \\ +\ 286 \\ \hline 1{,}325 \end{array}$$

Answer: 1,325

5 + 4 + 6 = 15.
Write 5 and carry 1 to the next column.
1 + 3 + 8 = 12.
Write 2 and carry 1 to the next column.
1 + 2 + 8 + 2 = 13.
Write 13.

PRACTICE 2 **Add: 197 + 385 + 46.**

Answer: 628

UNIT 4 Name ______________________ Date ____________

Horizontal and Vertical Addition of Larger Numbers

Add:

1. 485 + 693 1,178	2. 516 + 398 914	3. 963 + 847 1,810	4. 297 + 768 1,065	5. 642 + 579 1,221
6. 100 673 + 207 980	7. 1,071 341 + 5,001 6,413	8. 233 511 + 179 923	9. 4,444 1,234 + 5,561 11,239	10. 9,861 4,809 + 8,365 23,035
11. 587 926 + 180 1,693	12. 4,673 8,437 + 9,546 22,656	13. 796 284 + 325 1,405	14. 5,643 8,291 + 5,479 19,413	15. 8,654 4,766 + 3,919 17,339
16. 296 398 764 + 587 2,045	17. 831 935 247 + 472 2,485	18. 209 491 647 + 362 1,709	19. 625 287 536 + 169 1,617	20. 908 470 735 + 153 2,266
21. 924,408 635,297 753,317 + 146,819 2,459,841	22. 175,617 942,427 869,293 + 731,358 2,718,695	23. 836,935 718,361 196,502 + 850,624 2,602,422	24. 826,604 460,712 385,941 + 927,573 2,600,830	

Use Your Calculator to Add the Following:

25. 423 + 567 + 385 = 1,375
26. 98 + 483 + 956 + 85 = 1,622
27. 15 + 603 + 1,145 + 6,342 = 8,105
28. 376 + 493 + 102 + 315 = 1,286
29. 4,753 + 6,378 + 9,257 + 2,896 = 23,284
30. 974 + 65 + 376 + 487 + 598 + 88 = 2,588
31. 4,080 + 715 + 13,634 + 29 + 5 = 18,463
32. 127 + 646 + 30 + 29 = 832
33. 614,034 + 783,420 + 10,316 = 1,407,770

Business Applications:

34. A local restaurant served 40, 97, 132, 73, and 144 patrons over five days. Add up the number of patrons to determine the total.

486 patrons

35. A supermarket took inventory of some of its stock and found 10 cases of soup, 29 cases of canned vegetables, 57 cases of cat food, and 13 cases of frozen food. Add these amounts to determine the total number of cases.

109 cases

UNIT

5 Estimating Addition by Rounding

A quick estimate of the sum of a group of numbers can be found by rounding each number to its first digit and then adding these rounded numbers. Because we are only approximating an answer, this technique does not replace actual computations, but it is useful when having to make a decision based on rapid calculations.

EXAMPLE 1 **Estimate the sum of the following numbers by rounding each to its first digit and adding.**

		Rounded	
5,173,617		5,000,000	
5,198,482		5,000,000	
464,917		500,000	
325,061		300,000	
379,460		400,000	
291,117		300,000	
196,413		200,000	
160,053		200,000	
98,412		100,000	
93,615		90,000	
Answer: 12,381,147	actual sum	12,090,000	estimated sum

PRACTICE 1 **Estimate the sum of the following numbers by rounding each to its first digit and adding.**

57,945
55,405
53,924
46,741
43,887
39,221
29,316
24,323
\+ 10,628

Answer: *360,000*

EXAMPLE 2 **Estimate the sum by rounding each of the following numbers to the nearest hundred and adding.**

			Rounded	
	9,144		9,100	
	8,375		8,400	
	6,776		6,800	
	4,581		4,600	
	+ 2,319		+ 2,300	
Answer:	31,195	actual sum	31,200	estimated sum

PRACTICE 2 **Estimate the sum by rounding each of the following numbers to the nearest hundred and adding.**

	7,596
	6,813
	5,580
	5,488
	4,561
	4,500
	3,719
	+ 2,039
Answer:	40,300

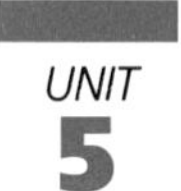

UNIT 5 Name ______________________ Date ______________

Estimating Addition by Rounding

Estimate the sum of each of the following groups of numbers by rounding each number to the given value and adding:

1. 658 + 831; hundred = 1,500
2. 6,721 + 9,344; thousand = 16,000
3. 823,093 + 951,176 + 285,615; thousand = 2,060,000
4. 917,554 + 816,659 + 674,419; thousand = 2,409,000
5. 31,481 + 29,563 + 15,141 + 19,938; ten thousand = 100,000
6. 656 + 315 + 708 + 190 + 314 + 818; 1st digit = 3,000
7. 908 + 470 + 735 + 591 + 930 + 491 + 619; 1st digit = 4,700
8. 6,538 + 6,783 + 2,645 + 125 + 985 + 27; 1st digit = 18,130

Business Applications:

9. Estimate the sum of the following maintenance costs by rounding each number to the first digit and adding: paint, $382; wallpaper, $57; windows, $637; and shrubbery; $53.

   $1,110

10. Estimate the total trading in all markets by rounding each number to the nearest hundred thousand: New York Exchange, 3,531,872; Midwest Exchange, 1,750,900; Boston Exchange, 1,157,359; and Philadelphia Exchange, 849,260.

    7,300,000

UNIT

6 Subtraction

Subtraction is the process of finding the difference between two numbers. The larger number being subtracted from is called the **minuend,** and the smaller number being subtracted is called the **subtrahend.** The result of the subtraction is called the **difference.**

To subtract one number from another:

1 Write the smaller number underneath the larger number, so numbers of the same place value are in the same column.

2 Subtract the smaller number (subtrahend) from the larger (minuend) in each column, and write the difference directly underneath.

3 To check the answer, add the difference and the subtrahend. If the original subtraction was correct, the total of these numbers must equal the minuend.

EXAMPLE 1 **Subtract: 936 – 123.**

$$\begin{array}{r} 936 \\ -\ 123 \\ \hline 813 \end{array}$$

Answer: 813

Check:
$$\begin{array}{r} 123 \\ +\ 813 \\ \hline 936 \end{array}$$

PRACTICE 1 **Subtract: 747 – 324.**

$$\begin{array}{r} 747 \\ -\ 324 \\ \hline \end{array}$$

Answer: 423

Check:

EXAMPLE 2 **Subtract: 596 – 481 and check.**

$$\begin{array}{r} 596 \\ -\ 481 \\ \hline 115 \end{array}$$

Answer: 115

Check:
$$\begin{array}{r} 481 \\ +\ 115 \\ \hline 596 \end{array}$$

PRACTICE 2 **Subtract: 818 – 603 and check.**

Check:

Answer: 215

UNIT 6 Name ____________________ Date ____________

Subtraction

Subtract and check the following:

1. 87 − 32 = 55
2. 31 − 20 = 11
3. 99 − 77 = 22
4. 65 − 43 = 22
5. 81 − 61 = 20
6. 125 − 114 = 11
7. 297 − 143 = 154
8. 369 − 257 = 112
9. 171 − 161 = 10
10. 963 − 722 = 241
11. 9,487 − 4,136 = 5,351
12. 8,371 − 6,151 = 2,220
13. 7,629 − 5,618 = 2,011
14. 1,820 − 1,710 = 110
15. 3,639 − 2,413 = 1,226
16. 32,798 − 11,694 = 21,104
17. 83,896 − 52,671 = 31,225
18. 65,099 − 24,099 = 41,000
19. 58,767 − 42,335 = 16,432
20. 97,873 − 15,360 = 82,513

21. 47 − 42 = 5
22. 428 − 416 = 12
23. 76 − 34 = 42
24. 93 − 41 = 52
25. 567 − 241 = 326
26. 690 − 570 = 120
27. 784 − 313 = 471
28. 965 − 362 = 603
29. 27,531 − 17,421 = 10,110
30. 87,424 − 63,223 = 24,201
31. 58,736 − 48,615 = 10,121
32. 98,449 − 33,227 = 65,222
33. 694,732 − 581,510 = 113,222
34. 834,949 − 724,918 = 110,031
35. 267,032 − 51,021 = 216,011
36. 389,576 − 20,002 = 369,574
37. 330,003 − 20,002 = 310,001
38. 416,164 − 16,164 = 400,000

Business Applications:

39. Subtract a down payment of $8,200 from the purchase price of $69,599 to determine the amount of loan needed.

$61,399

40. From an original inventory of 4,034 textbooks, subtract 3,021 textbooks sold to determine the bookstore's current inventory.

1,013 textbooks

UNIT

Subtraction with Borrowing

When the subtrahend (the bottom number) is larger than the minuend (the top number), **borrowing** is neccessary to find the difference. Sometimes repeated borrowing is necessary.

Another situation that involves borrowing occurs when zeros are part of the minuend, as in Example 3.

EXAMPLE 1 **Subtract: 737 − 419.**

$$\begin{array}{r} 7\overset{2}{\not3}{}^{1}7 \\ -\ 419 \\ \hline \end{array}$$

$$\begin{array}{r} 7\overset{2}{\not3}{}^{1}7 \\ -\ 419 \\ \hline 318 \end{array}$$

Answer: 318

1. Borrow because 9 cannot be subtracted from 7.
2. Borrow 1 ten from the 3.
3. Cross out 3 and make it a 2.
4. Add the borrowed 10 to 7 by writing a small 1 beside the seven, making it a 17.
5. Subtract the numbers in each column

PRACTICE 1 **Subtract: 592 − 317.**

Answer: 275

EXAMPLE 2 **Subtract: 3,452 − 678.**

$$\begin{array}{r} 3{,}4\overset{4}{\not5}{}^{1}2 \\ -\ \ 678 \\ \hline \end{array} \Rightarrow \begin{array}{r} 3{,}\overset{3}{\not4}{}^{1}\overset{4}{\not5}{}^{1}2 \\ -\ \ 678 \\ \hline 4 \end{array} \Rightarrow \begin{array}{r} \overset{2}{\not3}{,}{}^{1}\overset{3}{\not4}{}^{1}\overset{4}{\not5}{}^{1}2 \\ -\ \ 678 \\ \hline 74 \end{array} \Rightarrow \begin{array}{r} \overset{2}{\not3}{,}{}^{1}\overset{3}{\not4}{}^{1}\overset{4}{\not5}{}^{1}2 \\ -\ \ 678 \\ \hline 2{,}774 \end{array}$$

Answer: 2,774

1. Borrow 1 ten from the 5, making it 4, and add the 1 ten to the 2, making it 12.
2. Borrow 1 hundred from the 4, making it 3, and add it to the 4, making it 14.
3. Take 1 thousand from the 3, making it 2, and add it to the 3, making it 13.

PRACTICE 2 **Subtract: 6,715 − 937.**

Answer: 5,778

EXAMPLE 3 **Subtract: 8,007 − 649.**

$$\begin{array}{r} 8{,}007 \\ -\ \ 649 \\ \hline \end{array} \Rightarrow \begin{array}{r} \overset{7}{8},\overset{1}{0}07 \\ -\ \ 649 \\ \hline \end{array} \Rightarrow \begin{array}{r} \overset{7}{8},\overset{9}{\overset{1}{0}}\overset{1}{0}7 \\ -\ \ 649 \\ \hline \end{array} \Rightarrow \begin{array}{r} \overset{7}{8},\overset{9}{\overset{1}{0}}\overset{9}{\overset{1}{0}}\overset{1}{7} \\ -\ \ 649 \\ \hline \end{array} \Rightarrow \begin{array}{r} \overset{7}{8},\overset{9}{\overset{1}{0}}\overset{9}{\overset{1}{0}}\overset{1}{7} \\ -\ \ 649 \\ \hline 7{,}358 \end{array}$$

Answer: 7,358

1. Because there are zeros in both the tens and hundreds columns, borrow 1 thousand from the thousands column and write 10 in the hundreds column (10 hundreds equal 1 thousand).
2. Borrow 1 hundred from the hundreds column and write 10 in the tens column (10 tens equal 1 hundred).
3. Borrow 1 ten from the tens column and add it to the ones column, writing 17. Now subtract the numbers in each column.

PRACTICE 3 **Subtract: 4,005 − 768.**

Answer: 3,237

UNIT 7 Name ______________________ Date ______________

Subtraction with Borrowing

Subtract and check:

1. 406 − 154 = 252
2. 395 − 288 = 107
3. 623 − 515 = 108
4. 700 − 465 = 235
5. 912 − 453 = 459
6. 1,060 − 343 = 717
7. 2,360 − 1,272 = 1,088
8. 809 − 787 = 22
9. 1,120 − 744 = 376
10. 4,789 − 1,987 = 2,802

11. 854 − 578 = 276
12. 1,799 − 1,732 = 67
13. 8,641 − 1,904 = 6,737
14. 5,496 − 1,492 = 4,004
15. 1,584 − 920 = 664
16. 5,672 − 2,356 = 3,316
17. 74,760 − 39,817 = 34,943
18. 52,365 − 15,423 = 36,942
19. 78,567 − 32,782 = 45,785
20. 38,607 − 3,867 = 34,740
21. 7,040 − 3,094 = 3,946
22. 6,007 − 3,628 = 2,379

Business Applications:

23. Determine the new selling price of a calculator by subtracting a discount of $19 from its original selling price of $33.

$14

24. Determine the current balance in a checkbook by subtracting a check in the amount of $197 from the original balance of $331.

$134

UNIT 8 How to Dissect and Solve Word Problems: Addition and Subtraction

The frequent use of arithmetic in business makes it necessary to apply the right arithmetic processes to effectively solve problems. Effective problem solving depends on the ability to persevere at organizing and properly analyzing the information presented in a problem and calculating answers correctly. Some general suggestions for problem solving are:

1. Read and reread the problem to clearly understand what it says.
2. Gather the appropriate facts needed to solve the problem.
3. Decide what you need to know or what you need to be able to calculate before solving the problem.
4. Be sure you understand how to carry out the necessary calculations.
5. Perform whatever procedures are necessary to solve the problem.
6. Decide on the reasonableness of your answer, and if at all possible, check the answer.
7. State the answer, including the correct unit of measure.

The chart below summarizes this information and can be used repeatedly until you have mastered problem solving.

Gather the facts.	What am I solving for?	What must I need to know or calculate before solving the problem?	Key points to remember.

EXAMPLE 1 **A summary of voter turnout by precinct in a recent city election revealed that there were 6,671 votes cast in Precinct One; 4,728 votes cast in Precinct Two; 2,839 votes cast in Precinct Three; and 2,255 votes cast in Precinct Four. What was the total vote?**

Gather the facts.	What am I solving for?	What must I need to know or calculate before solving the problem?	Key points to remember.
Precinct 1 = 6,671 Precinct 2 = 4,728 Precinct 3 = 2,839 Precinct 4 = 2,255	Total vote	Precincts 1 + 2 + 3 + 4 = Total vote	Line numbers up vertically. Add numbers in each column from top to bottom. Check by adding from bottom to top.

$$\begin{array}{r} 6{,}671 \\ 4{,}728 \\ 2{,}839 \\ +\ 2{,}255 \\ \hline 16{,}493 \end{array}$$

Answer: 16,493 votes

PRACTICE 1 **If the area of Maine is approximately 30,000 square miles; New Hampshire, 9,280; Vermont, 9,056; Massachusetts, 7,800; Rhode Island, 1,306; and Connecticut, 4,674, what is the total square mileage of this six state New England region?**

Answer: 62,116 square miles

EXAMPLE 2 **Essex Press originally printed 3,550 copies of a real estate manual. In January, they distributed 375 copies; in March, 278 copies; in June, 560 copies; and in August, 349 copies. How many copies are left from the original printing?**

Gather the facts.	What am I solving for?	What must I need to know or calculate before solving the problem?	Key points to remember.
Original copies = 3,550 Distributed copies = 375 278 560 349	Total copies distributed Total copies left over	Total printed – Total distributed = Total copies left	Minuend – Subtranend = Difference.

$375 + 278 + 560 + 349 = 1{,}562$ Total copies distributed

$$\begin{array}{r} 3{,}550 \\ -\ 1{,}562 \\ \hline 1{,}988 \end{array}$$

Answer: 1,988 copies left

PRACTICE 2 **The Gardner family has a monthly budget of $1,600. If they pay $882 toward their mortgage, $125 for utilities, and $180 for food, how much do they have left for other expenses?**

Answer: $413

UNIT 8 Name ______________________ Date ____________

How to Dissect and Solve Word Problems: Addition and Subtraction

Solve the following problems:

1. A local realtor selling four housing lots estimates the size of each to be: 5,725 square feet; 6,087 square feet; 4,325 square feet; and 6,654 square feet. What is the total square footage of the four lots?

   22,791 square feet

2. Professor Slater's classes are held in classrooms that can accommodate 25, 29, 34, 36, 150, 27, and 72 students. What is the total seating capacity of these classrooms?

   373 seats

3. Attendance at local football games one Saturday in November was as follows: 2,237 people at Holt Stadium; 1,763 people at Sims Stadium; 3,096 at Veterans Field; and 1,622 at Dillon Field. What was the total attendance at these games?

   8,718 people

4. During the month of October, Diane recorded the following earnings from her window treatment business: $1,725, $950, $625, $135, $77, and $24. What did she earn altogether?

   $3,536

5. The population of Belleville is 17,097 and the population of Kearny is 22,378. By what amount is the population of Kearny greater than that of Belleville?

   5,281 people

6. In a walkathon at St. Mary's School, the two fourth-grade classes completed a total of 167 and 191 laps around the school's athletic field. The two sixth-grade classes completed a total of 217 and 153 laps. How many more laps did the sixth graders complete than did the fourth graders?

   12 laps

7. Colleen stuffed 23 trash bags with leaves she raked in her backyard. Her sister stuffed 14 trash bags with leaves she had raked. How many more bags of leaves did Colleen rake?

   9 bags

8. Michael's bowling scores were 89, 93, and 97 for three strings. His younger brother Mark's scores were 64, 71, and 67. By what amount did Michael's scores exceed Mark's?

   77 points

9. A 16-foot boat has a capacity of 1,500 pounds. If the weights of the six people in the craft are 193 pounds, 105 pounds, 57 pounds, 73 pounds, 169 pounds, and 112 pounds, how much more weight can the boat hold?

   791 pounds

10. The Tobin's family car had an odometer reading of 33,742 miles at the start of their vacation. At vacation's end, the odometer read 34,895 miles. How many miles did they travel on the vacation?

    1,153 miles

11. Three phone operators at a telethon received 670 phone calls, 748 phone calls, and 590 phone calls over five days. A total of 3,502 phone calls were received. What is the difference between the total number of phone calls and the total phone calls taken by the three operators?

1,494 calls

12. An associates degree in respiratory therapy at West Shore Community College requires the completion of 72 credits. If Rodney H. has completed 17, 13, 15, and 6 credits thus far, how many more credits must he complete to receive a degree?

21 credits

13. Joe received two estimates for building a deck in his backyard; one for $1,475 and a second for $1,850. What was the difference between the two estimates?

$375

14. Paul and Rose calculated the total ticket sales for a recent dance party to be $1,657. If the cost of the band was $400; food, $352; the facility, $600; and miscellaneous expenses, $276, what profit, if any, did Paul and Rose make?

$29

15. Attendance at three home basketball games was 12,076; 15,909; and 13,173. Attendance at three away games was 9,874; 10,517; and 14,319. By what margin did home attendance exceed attendance at away games?

6,448 people

16. Bill's annual income is $62,599. His wife Sharon's annual income is $81,900. If they estimate their total annual expenses to be $60,480, what is the difference between their combined incomes and total expenses?

$84,019

17. Rhonda, Gail, Diane, and Julie spend $478, $350, $316, and $529 respectively on Christmas gifts for family and friends. What was the total amount of money spent on gifts?

$1,673

18. Juan estimates that he spent $3,975 last year driving his own car to work. Trying to save money, he now uses public transportation and estimates he has spent $2,770 this year. How much money has he saved by using public transportation?

$1,205

19. Youth soccer programs in three cities enroll 3,217; 1,819; and 2,425 children. The Little League baseball programs in each of the same towns enroll 2,920; 2,076; and 1,965 children. Which set of programs enrolls more children; soccer or baseball and by how many children?

soccer; 500 children

20. Judy finds a car she can buy for $1,603 less than the cost of the car Mike has found, which costs $11,693. What is the price of the car Judy found?

$10,090

UNIT

9 Multiplication—Basic Facts

Multiplication is a quick way of adding a number a repeated number of times. Numbers that are multiplied are called **factors.** When arranged vertically the top number is called the **multiplicand** and the bottom number is the **multiplier.** The result of the multiplication process is called the **product.**

One of two symbols is used to express multiplication. The **times sign,** ×, and the **round dot,** •, when placed between numbers indicate the numbers are to be multiplied.

Knowledge of basic multiplication facts is essential for completing longer multiplication problems later. Completing the charts below will aid this process.

UNIT 9 Name ______________________ Date ______________

Multiplication—Basic Facts

Fill in the blanks by multiplying each number in the top row by those in the left-hand column:

×	**1**	**2**	**3**	**4**	**5**	**6**	**7**	**8**	**9**	**10**	**11**	**12**
1	1	2	3	4	5	6	7	8	9	10	11	12
2	2	4	6	8	10	12	14	16	18	20	22	24
3	3	6	9	12	15	18	21	24	27	30	33	36
4	4	8	12	16	20	24	28	32	36	40	44	48
5	5	10	15	20	25	30	35	40	45	50	55	60
6	6	12	18	24	30	36	42	48	54	60	66	72
7	7	14	21	28	35	42	49	56	63	70	77	84
8	8	16	24	32	40	48	56	64	72	80	88	96
9	9	18	27	36	45	54	63	72	81	90	99	108
10	10	20	30	40	50	60	70	80	90	100	110	120
11	11	22	33	44	55	66	77	88	99	110	121	132
12	12	24	36	48	60	72	84	96	108	120	132	144

Fill in the blanks by multiplying each number in the top row by those in the left-hand column:

×	7	4	1	3	11	9	5	12	8	2	6	10
8	56	32	8	24	88	72	40	96	64	16	48	80
4	28	16	4	12	44	36	20	48	32	8	24	40
11	77	44	11	33	121	99	55	132	88	22	66	110
6	42	24	6	18	66	54	30	72	48	12	36	60
9	63	36	9	27	99	81	45	108	72	18	54	90
12	84	48	12	36	132	108	60	144	96	24	72	120
1	7	4	1	3	11	9	5	12	8	2	6	10
3	21	12	3	9	33	27	15	36	24	6	18	30
10	70	40	10	30	110	90	50	120	80	20	60	100
2	14	8	2	6	22	18	10	24	16	4	12	20
7	49	28	7	21	77	63	35	84	56	14	42	70
5	35	20	5	15	55	45	25	60	40	10	30	50

UNIT

10 Multiplication of Larger Numbers

When multiplying larger numbers, the numbers obtained and arranged between the multiplier and the product are called **partial products.** To multiply larger numbers:

1. Line the multiplier (bottom number) up with the multiplicand (top number), so the units stand under units, tens under tens, and so on. It is usually best to make the smaller number the multiplier.
2. Multiply the right-hand digit of the multiplier times the right-hand digit in the multiplicand.
3. Write down the right-hand digit of the product, starting directly underneath the multiplying digit.
4. Carry any left-hand digits of the product to the next number in the multiplicand and add it to the next product.
5. Keep multiplying in this fashion, moving left through all the numbers in the multiplicand.
6. Move left through the multiplier, multiplying each digit of the multiplicand by each digit of the multiplier. The right-hand digit of each partial product is written directly underneath the multiplying digit.
7. Add the partial products from right to left to obtain the final product.

EXAMPLE 1 **Multiply: 37 × 4.**

$$\begin{array}{r} {}^{2}37 \\ \times \quad 4 \\ \hline 8 \end{array}$$

1. Multiply 4 × 7 (4 × 7 = 28).
2. Write down 8 and carry 2 to the next number.

$$\begin{array}{r} {}^{2}37 \\ \times \quad 4 \\ \hline 148 \end{array}$$

1. Multiply 4 × 3 (4 × 3 = 12).
2. Add 12 + 2 (12 + 2 = 14).
3. Write down 14.

Answer: 148

PRACTICE 1 **Multiply: 26 × 3.**

Answer: 78

EXAMPLE 2 **Multiply: 439 × 82.**

$$\begin{array}{r} {}^{1} \\ 439 \\ \times \quad 82 \\ \hline 878 \end{array}$$

1. Multiply by 2.

$$\begin{array}{r} {}^{3\,7} \\ 439 \\ \times \quad 82 \\ \hline 878 \\ 3{,}512 \\ \hline 35{,}998 \end{array}$$

2. Multiply by 8.
3. Add the partial products to obtain the product 35,998.

Answer: 35,998

PRACTICE 2 **Multiply: 316 × 53.**

Answer: 16,748

UNIT 10 Name ______________________ Date ______________

Multiplication of Larger Numbers

Multiply the following:

1. 76 × 4 304	2. 89 × 6 534	3. 56 × 2 112	4. 27 × 3 81	5. 21 × 8 168
6. 53 × 74 3,922	7. 84 × 49 4,116	8. 58 × 63 3,654	9. 28 × 57 1,596	10. 68 × 43 2,924
11. 99 × 25 2,475	12. 85 × 49 4,165	13. 17 × 91 1,547	14. 48 × 62 2,976	15. 94 × 13 1,222
16. 578 × 42 24,276	17. 996 × 87 86,652	18. 234 × 67 15,678	19. 625 × 39 24,375	20. 716 × 46 32,936
21. 268 × 341 91,388	22. 722 × 317 228,874	23. 623 × 542 337,666	24. 822 × 114 93,708	25. 314 × 264 82,896

Business Applications:

26. Multiply to calculate the total rent expense for 6 months if the space rented costs $650 per month.

$3,900

27. Multiply to determine the total number of miles one can drive on one tank of gas if the tank holds 14 gallons and the car gets 27 miles per gallon.

378 miles

Multiplication Involving Zeros

Multiplication when zeros are present is actually quite simple. Always remember that zero times anything and anything times zero always equal zero.

To multiply numbers ending in zeros:

1 When zeros are at the end of the multiplicand or the multiplier, or both, disregard the zeros and multiply the nonzero digits.

2 Count the number of zeros in the problem and add this number of zeros to the answer obtained above.

EXAMPLE 1 **Multiply: 3,076 × 402.**

```
   3,076              3,076
 ×   402            ×   402
   6,152     or       6,152
   0000              12,304
  12,304          1,236,552
1,236,552
```

Answer: 1,236,552

1. Multiply by 2.
2. Multiply by 0.
3. Multiply by 4.
4. Add the partial products.

PRACTICE 1 **Multiply: 4,035 × 606.**

```
  4,035
×   606
```

Answer: 2,445,210

EXAMPLE 2 **Multiply: 376,000 × 470.**

1. Multiply:

```
   376
 ×  47
 2,632
1,504
17,672
```

Answer: 2. Adding 4 zeros gives 176,720,000.

PRACTICE 2 **Multiply: 416,000 × 3,200.**

Answer: 1,331,200,000

EXAMPLE 3 **Multiply:**

$77 \times 10 = 770$
$77 \times 100 = 7{,}700$
$77 \times 1{,}000 = 77{,}000$
$77 \times 20 = 1{,}540$
$77 \times 200 = 15{,}400$
$77 \times 2{,}000 = 154{,}000$
$77 \times 30 = 2{,}310$
$77 \times 300 = 23{,}100$
$77 \times 3{,}000 = 231{,}000$

PRACTICE 3 **Multiply:**

$89 \times 10 =$ 890
$89 \times 100 =$ 8,900
$89 \times 1{,}000 =$ 89,000
$89 \times 20 =$ 1,780
$89 \times 200 =$ 17,800
$89 \times 2{,}000 =$ 178,000
$89 \times 30 =$ 2,670
$89 \times 300 =$ 26,700
$89 \times 3{,}000 =$ 267,000

UNIT **11** Name ______________________ Date ______________

Multiplication Involving Zeros

Multiply the following:

1. 623 × 504 =

   313,992

2. 431 × 208 =

   89,648

3. 822 × 702 =

   577,044

4. 3,206 × 105 =

   336,630

5. 4,003 × 402 =

   1,609,206

6. 5,204 × 401 =

   2,086,804

7. 7,003 × 1,008 =

   7,059,024

8. 5,009 × 3,007 =

   15,062,063

9. 6,203 × 5,006 =

   31,052,218

10. 4,022 × 6,009 =

    24,168,198

11. 264 × 3,800 =

    1,003,200

12. 314 × 1,300 =

    408,200

13. 232 × 4,100 =

    951,200

14. 422 × 2,300 =

    970,600

15. 510 × 3,300 =

    1,683,000

16. 420 × 5,300 =

    2,226,000

17. 24,000 × 360 =

    8,640,000

18. 52,000 × 180 =

    9,360,000

19. 146,000 × 3,600 =

    525,600,000

20. 729,000 × 47,000 =

    34,263,000,000

21. 5,250,000 × 47,000 =

    246,750,000,000

22. 3,300 × 100 =

330,000

23. 416 × 2,000 =

832,000

24. 9,729 × 10,000 =

97,290,000

25. 4,681 × 10,000 =

46,810,000

26. 1,739 × 40 =

69,560

27. 3,794 × 400 =

1,517,600

28. 2,975 × 100 =

297,500

29. 2,975 × 1,000 =

2,975,000

30. 2,975 × 10,000 =

29,750,000

31. 2,975 × 50 =

148,750

32. 2,975 × 500 =

1,487,500

33. 2,975 × 5,000 =

14,875,000

34. 7,876 × 60 =

472,560

35. 7,876 × 600 =

4,725,600

36. 7,876 × 6,000 =

47,256,000

Business Applications:

37. Multiply to determine the total cost of 150 tickets costing $14 per ticket.

$2,100

38. Multiply to calculate the total number of golf balls that can be manufactured over 15 work shifts if 3,000 balls can be manufactured per shift.

45,000 balls

UNIT

12 Division of Whole Numbers—Basic Facts

The operation of division is the opposite of multiplication and determines how many times a given number is contained in another as a multiple. Division may be indicated by the division sign ÷, or the notation $\overline{)}$. The number to be divided is called the **dividend.** The number by which we divide is called the **divisor** and the result obtained by the division is called the **quotient.** For instance, in the examples

$\begin{array}{r} 3 \\ 4\overline{)12} \end{array}$ or $12 \div 4 = 3$,

the dividend is 12, the divisor is 4, and the quotient is 3. To prove this division problem we would multiply the divisor times the quotient to get the dividend.

Some important division facts are:

1 Zero divided by any number always equals zero.

Example: $\begin{array}{r} 0 \\ 4\overline{)\,0} \end{array}$

2 Division by zero is impossible.

3 Any number divided by itself always equals one.

Example: $\begin{array}{r} 1 \\ 4\overline{)\,4} \end{array}$

4 Any number divided by one is the number itself.

Example: $\begin{array}{r} 4 \\ 1\overline{)\,4} \end{array}$

Because division is the opposite of multiplication, swift and accurate completion of the following exercises is dependent on one's mastery of multiplication facts. Complete the following exercises as quickly as possible. Review any multiplication facts as necessary.

UNIT 12 Name ______________________ Date ______________

Division of Whole Numbers—Basic Facts

Divide:

1. $\begin{array}{r} 3 \\ 7\overline{)21} \end{array}$ $\begin{array}{r} 1 \\ 3\overline{)3} \end{array}$ $\begin{array}{r} 9 \\ 6\overline{)54} \end{array}$ $\begin{array}{r} 4 \\ 8\overline{)32} \end{array}$ $\begin{array}{r} 2 \\ 4\overline{)8} \end{array}$ $\begin{array}{r} 5 \\ 9\overline{)45} \end{array}$ $\begin{array}{r} 6 \\ 4\overline{)24} \end{array}$ $\begin{array}{r} 8 \\ 5\overline{)40} \end{array}$

2. $\begin{array}{r} 2 \\ 3\overline{)6} \end{array}$ $\begin{array}{r} 7 \\ 8\overline{)56} \end{array}$ $\begin{array}{r} 4 \\ 4\overline{)16} \end{array}$ $\begin{array}{r} 5 \\ 7\overline{)35} \end{array}$ $\begin{array}{r} 1 \\ 8\overline{)8} \end{array}$ $\begin{array}{r} 6 \\ 9\overline{)54} \end{array}$ $\begin{array}{r} 8 \\ 3\overline{)24} \end{array}$ $\begin{array}{r} 2 \\ 8\overline{)16} \end{array}$

3.	$\begin{array}{r}8\\7\overline{)56}\end{array}$	$\begin{array}{r}1\\5\overline{)5}\end{array}$	$\begin{array}{r}2\\9\overline{)18}\end{array}$	$\begin{array}{r}3\\9\overline{)27}\end{array}$	$\begin{array}{r}4\\5\overline{)20}\end{array}$	$\begin{array}{r}5\\6\overline{)30}\end{array}$	$\begin{array}{r}6\\3\overline{)18}\end{array}$	$\begin{array}{r}7\\9\overline{)63}\end{array}$
4.	$\begin{array}{r}9\\2\overline{)18}\end{array}$	$\begin{array}{r}5\\2\overline{)10}\end{array}$	$\begin{array}{r}3\\6\overline{)18}\end{array}$	$\begin{array}{r}4\\3\overline{)12}\end{array}$	$\begin{array}{r}4\\7\overline{)28}\end{array}$	$\begin{array}{r}5\\3\overline{)15}\end{array}$	$\begin{array}{r}6\\1\overline{)6}\end{array}$	$\begin{array}{r}6\\6\overline{)36}\end{array}$
5.	$\begin{array}{r}7\\6\overline{)42}\end{array}$	$\begin{array}{r}7\\4\overline{)28}\end{array}$	$\begin{array}{r}1\\6\overline{)6}\end{array}$	$\begin{array}{r}2\\7\overline{)14}\end{array}$	$\begin{array}{r}5\\8\overline{)40}\end{array}$	$\begin{array}{r}9\\4\overline{)36}\end{array}$	$\begin{array}{r}9\\8\overline{)72}\end{array}$	$\begin{array}{r}7\\3\overline{)21}\end{array}$
6.	$\begin{array}{r}3\\3\overline{)9}\end{array}$	$\begin{array}{r}9\\3\overline{)27}\end{array}$	$\begin{array}{r}2\\5\overline{)10}\end{array}$	$\begin{array}{r}8\\1\overline{)8}\end{array}$	$\begin{array}{r}7\\5\overline{)35}\end{array}$	$\begin{array}{r}1\\4\overline{)4}\end{array}$	$\begin{array}{r}5\\5\overline{)25}\end{array}$	$\begin{array}{r}6\\7\overline{)42}\end{array}$
7.	$\begin{array}{r}9\\7\overline{)63}\end{array}$	$\begin{array}{r}3\\5\overline{)15}\end{array}$	$\begin{array}{r}1\\7\overline{)7}\end{array}$	$\begin{array}{r}7\\7\overline{)49}\end{array}$	$\begin{array}{r}4\\2\overline{)8}\end{array}$	$\begin{array}{r}9\\1\overline{)9}\end{array}$	$\begin{array}{r}6\\8\overline{)48}\end{array}$	$\begin{array}{r}2\\6\overline{)12}\end{array}$
8.	$\begin{array}{r}1\\2\overline{)2}\end{array}$	$\begin{array}{r}5\\4\overline{)20}\end{array}$	$\begin{array}{r}9\\9\overline{)81}\end{array}$	$\begin{array}{r}7\\2\overline{)14}\end{array}$	$\begin{array}{r}1\\9\overline{)9}\end{array}$	$\begin{array}{r}3\\2\overline{)6}\end{array}$	$\begin{array}{r}9\\5\overline{)45}\end{array}$	$\begin{array}{r}8\\6\overline{)48}\end{array}$

Business Applications:

9. Divide to determine the average number of parts a machine operator can produce per hour if 81 parts are produced over 9 hours.

   9 parts

10. Divide to determine the average number of hours 7 employees worked if the total number of hours worked was 49.

    7 hours

UNIT 13 Short Division

Short division is a process where the multiplication and subtraction steps involved in the division process are not written down. Short division is used when the divisor is small enough to allow one to mentally do the multiplication and subtraction.

1. Write the divisor to the left of the dividend with the notation $\overline{)\quad}$ between them.
2. Beginning with the left-hand digit, divide each digit of the dividend by the divisor, writing the quotient directly above the dividend.
3. If there is a remainder, "carry it over" to the next digit in the dividend and divide this newly formed number by the divisor.
4. If any number in the dividend is smaller than the divisor, write a zero above it and carry this number to the next digit. Divide this newly formed number by the divisor.
5. If a remainder exists after the last digit, place it to the right of the quotient and precede it with the letter *R*.

EXAMPLE 1 **Divide: 2,884 ÷ 4.**

Answer:

$$\begin{array}{r} 721 \\ 4\overline{)2{,}884} \end{array}$$

Rewrite the problem.

1. Because 2 can't be divided by 4, we carry it to the 8. Divide 28 by 4 (28 ÷ 4 = 7). Place 7 above 8.
2. Divide 8 by 4 (8 ÷ 4 = 2). Place 2 above 8.
3. Divide 4 by 4 (4 ÷ 4 = 1). Place 1 above 4.

PRACTICE 1 **Divide: 2,469 ÷ 3.**

Answer: 823

EXAMPLE 2 **Divide: 8,430 ÷ 6.**

Answer:

$$\begin{array}{r} 1{,}405 \\ 6\overline{)8{,}430} \end{array}$$

Rewrite the problem.

1. Divide 8 by 6 (8 ÷ 6 = 1 with a remainder of 2). Place 1 above 8 and carry 2 to 4 making it 24.
2. Divide 24 by 6 (24 ÷ 6 = 4). Place 4 above 4.
3. Because 3 is smaller than the divisor 6, place 0 above 3 and carry 3 to 0, making it 30.
4. Divide 30 by 6 (30 ÷ 6 = 5). Place 5 above 0.

PRACTICE 2 **Divide: 7,545 ÷ 5.**

Answer: 1,509

UNIT **13** Name ______________________ Date ______________

Short Division

Divide:

1. 4,862 ÷ 2 = 2,431
2. 48,844 ÷ 4 = 12,211
3. 9,963 ÷ 3 = 3,321
4. 5,555 ÷ 5 = 1,111
5. 68,242 ÷ 2 = 34,121
6. 66,666 ÷ 6 = 11,111
7. 36,488 ÷ 4 = 9,122
8. 72,999 ÷ 9 = 8,111
9. 2,177 ÷ 7 = 311
10. 3,228 ÷ 4 = 807
11. 28,357 ÷ 7 = 4,051
12. 40,525 ÷ 5 = 8,105
13. 36,426 ÷ 6 = 6,071
14. 184,210 ÷ 2 = 92,105
15. 85,688 ÷ 8 = 10,711
16. 273,615 ÷ 3 = 91,205
17. 16,484 ÷ 4 = 4,121
18. 510,384 ÷ 7 = 72,912
19. 36,426 ÷ 6 = 6,071
20. 65,172 ÷ 7 = 9,310 R2
21. 2,926 ÷ 9 = 325 R1
22. 6,738 ÷ 8 = 842 R2
23. 5,677 ÷ 5 = 1,135 R2
24. 7,139 ÷ 4 = 1,784 R3
25. 67,359 ÷ 8 = 8,419 R7
26. 33,375 ÷ 6 = 5,562 R3
27. 70,588 ÷ 8 = 8,823 R4
28. 63,027 ÷ 7 = 9,003 R6

Business Applications:

29. Over 7 days, 1,764 gallons of milk are sold at a local supermarket. Divide to calculate the average number of gallons sold each day.

    252 gallons

30. The total payroll for a small office of 9 workers is $292,500. Divide to determine the average salary of each worker.

    $32,500

UNIT

14 Long Division

Long division is the division process where the multiplication and subtraction operations are written down. The process is used when the divisor contains two or more digits.

1. Write the divisor to the left of the dividend.
2. Divide the smallest number of digits in the dividend that contains the divisor one or more times. Write this quotient above the digit farthest to the right in the number divided.
3. Multiply the divisor by this quotient and subtract the product from the portion of the dividend used.
4. Bring down the next digit of the dividend to this remainder.
5. Continue dividing until all figures of the dividend have been brought down and divided.
6. If any portion of the dividend doesn't contain the divisor, place a zero in the quotient, bring down the next digit, and divide as before.

EXAMPLE 1 **Divide: 17,158 ÷ 23.**

Answer:

```
      746
23)17,158
   161
   ---
    105
     92
    ---
     138
     138
     ---
       0
```

Rewrite the problem.

1. The smallest number divisible by 23 is 171.
   Divide 171 by 23 (171 ÷ 23 = 7).
   Place 7 above 1.
   Multiply 7 × 23 (7 × 23 = 161).
   Subtract 171 − 161 (171 − 161 = 10).
2. Bring down 5.
   Divide 105 by 23 (105 ÷ 23 = 4).
   Place 4 above 5.
   Multiply 4 × 23 (4 × 23 = 92).
   Subtract 105 − 92 (105 − 92 = 13).
3. Bring down 8.
   Divide 138 by 23 (138 ÷ 23 = 6).
   Place 6 above 8.
   Multiply 6 × 23 (6 × 23 = 138).
   Subtract 138 − 138 (138 − 138 = 0).
4. No remainder exists.

PRACTICE 1 **Divide: 67,536 ÷ 12.**

Answer: 5,628

EXAMPLE 2 **Divide: 36,838 ÷ 18.**

Answer:

```
      2,046 R10
18 ) 36,838
     36
     --
      08
      00
      --
       83
       72
       --
       118
       108
       ---
        10
```

Rewrite the problem.

1. The smallest number divisible by 18 is 36.
   Divide 36 by 18 (36 ÷ 18 = 2).
   Place 2 above 6.
   Multiply 2 × 18 (2 × 18 = 36).
   Subtract 36 − 36 (36 − 36 = 0).
2. Bring down 8.
   Divide 8 by 18 (8 ÷ 18 = 0).
   Place 0 above 8.
   Multiply 0 × 18 (0 × 18 = 0).
   Subtract 8 − 0 (8 − 0 = 8).
3. Bring down 3.
   Divide 83 by 18 (83 ÷ 18 = 4).
   Place 4 above 3.
   Multiply 4 × 18 (4 × 18 = 72).
   Subtract 83 − 72 (83 − 72 = 11).
4. Bring down 8.
   Divide 118 by 18 (118 ÷ 18 = 6).
   Place 6 above 8.
   Multiply 6 × 18 (6 × 18 = 108).
   Subtract 118 − 108 (118 − 108 = 10).
5. 10 is the remainder.

PRACTICE 2 **Divide: 79,368 ÷ 36.**

Answer: 2,204 R24

UNIT **14** Name ______________________ Date ______________

Long Division

Divide:

1. 178,464 ÷ 16 =

   11,154

2. 15,341 ÷ 29 =

   529

3. 463,554 ÷ 39 =

   11,886

4. 1,299,123 ÷ 17 =

   76,419

5. 161,700 ÷ 15 =

   10,780

6. 47,653 ÷ 24 =

   1,985 R13

7. 765,431 ÷ 42 =

   18,224 R23

8. 6,783 ÷ 15 =

   452 R3

9. 7,831 ÷ 18 =

   435 R1

10. 9,767 ÷ 22 =

    443 R21

11. 7,654 ÷ 24 =

    318 R22

12. 767,500 ÷ 23=

    33,369 R13

13. 250,765 ÷ 34 =

    7,375 R15

14. 5,571,489 ÷ 43 =

    129,569 R22

15. 153,598 ÷ 29 =

    5,296 R14

16. 301,147 ÷ 63 =

    4,780 R7

17. 40,231 ÷ 75 =

    536 R31

18. 52,761,878 ÷ 126 =

    418,745 R8

19. $92{,}550 \div 25 =$

3,702

20. $7{,}461{,}300 \div 95 =$

78,540

21. $1{,}893{,}312 \div 912 =$

2,076

22. $833{,}382 \div 207 =$

4,026

23. $52{,}847{,}241 \div 607 =$

87,063

24. $946{,}656 \div 1{,}038 =$

912

25. $1{,}193{,}288 \div 45 =$

26,517 R23

26. $5{,}973{,}467 \div 243 =$

24,582 R41

27. $69{,}372{,}168 \div 342 =$

202,842 R204

Business Applications:

28. Divide $8,500 by 17 months to determine the monthly payment one must make to pay off this amount.

$500

29. Divide to determine a sales professional's average monthly salary over 16 months if the total earnings over this time were $616,000.

$38,500

UNIT

15 Division with Numbers Ending in Zero

The following steps save a great deal of time when performing division with numbers ending in zero.

When the dividend and divisor both end in zeros:

1. Count the number of ending zeros in the divisor.
2. Drop the same number of zeros in the dividend as in the divisor, counting from right to left.
3. Divide the remaining digits of the dividend by the remaining digits of the divisor.
4. The result is the quotient.

When only the divisor contains zeros:

1. Count the number of ending zeros in the divisor.
2. Drop the same number of digits in the dividend as there are ending zeros in the divisor, counting from right to left.
3. Divide the remaining digits of the dividend by the remaining digits of the divisor.
4. If no remainder exists after this division, the number obtained is the quotient, and the digits dropped from the dividend represent the remainder.
5. If a remainder does exist after this division, the number obtained is the quotient, and the remainder is written to the left of the digits dropped from the divisor. This newly formed number represents the remainder.

EXAMPLE 1 **Divide: 75,000 ÷ 1,000.**

$75 \div 1 = 75$ Drop three zeros from the dividend and divisor.

Answer: 75

PRACTICE 1 **Divide: 365,000 ÷ 1,000.**

Answer: 365

EXAMPLE 2 **Divide: 7,439,607 ÷ 600.**

$74{,}396 \div 6 =$

$$6\overline{)74{,}396}\quad 12{,}399 \text{ R}207$$

1. Drop two zeros from the divisor and two digits from the dividend.
2. Divide 74,396 by 6.
3. Write the remainder 2 to the left of the dropped digits 07 to form the remainder 207.

Answer: 12,399 R207

PRACTICE 2 **Divide: 34,716 ÷ 900.**

Answer: 38 R516

UNIT Name ______________________ Date ____________

15 Division with Numbers Ending in Zero

Divide the following using the methods described in this unit:

1. 6,500 ÷ 10 =

 650

2. 65,000 ÷ 100 =

 650

3. 65,000 ÷ 1,000 =

 65

4. 375,000 ÷ 1,000 =

 375

5. 375,000 ÷ 100 =

 3,750

6. 375,000 ÷ 10 =

 37,500

7. 50,670 ÷ 100 =

 506 R70

8. 320,762 ÷ 1,000 =

 320 R762

9. 14,030,731 ÷ 10,000 =

 1,403 R731

10. 6,080 ÷ 1,000 =

 6 R80

11. 12,500 ÷ 1,000 =

 12 R500

12. 9,021,300,640 ÷ 100,000 =

 90,213 R640

13. 437,661 ÷ 800 =

 547 R61

14. 46,820 ÷ 400 =

 117 R20

15. 130,626 ÷ 800 =

 163 R226

16. 76,173 ÷ 320 =

 238 R13

17. 378,000 ÷ 1,200 =

 315

18. 674,321 ÷ 11,200 =

 60 R2,321

Business Applications:

19. Determine the number of monthly payments needed to pay off a $300,000 loan by dividing this amount by the monthly payment of $500.

 600 payments

20. Divide to determine the average yearly amount that must be saved each year for the next 10 years if at the end of this time $89,000 must be set aside for college tuition.

 $8,900

UNIT

16 How to Dissect and Solve Word Problems: Multiplication and Division

The chart that follows serves as a guide for dissecting and solving the word problems in this unit. Review, if necessary, the general suggestions for solving word problems in Unit 8.

Gather the facts.	What am I solving for?	What must I need to know or calculate before solving the problem?	Key points to remember.

EXAMPLE 1 **Sal purchased 5 boxes of donuts each containing 13 donuts. How many donuts were purchased?**

Gather the facts.	What am I solving for?	What must I need to know or calculate before solving the problem?	Key points to remember.
5 boxes 13 donuts per box	Total number of donuts	Donuts per box × Number of boxes = Total number of donuts	Multiplication is repeated addition.

Answer: $13 \times 5 = 65$ donuts

PRACTICE 1 **Ed buys 3 boxes of playing cards for a casino night fundraiser. If each box contains 8 decks of cards, what is the total number of decks purchased?**

Answer: 24 decks

EXAMPLE 2 **A delivery of 192 boxes of manila folders was made to the administrative offices of a small college. If the boxes of folders were to be distributed equally to 16 offices, how many did each receive?**

Gather the facts.	What am I solving for?	What must I need to know or calculate before solving the problem?	Key points to remember.
192 boxes 16 offices	Number of boxes per office	Boxes ÷ Offices = Number of boxes per office	Division is the reverse of multiplication.

Answer: 192 ÷ 16 = 12 boxes

PRACTICE 2 **Brendan estimates that it will take 68 minutes to collect the money owed him by the 17 households to which he delivers newspapers. How much time can he spend at each house?**

Answer: 4 minutes

UNIT Name ______________________ Date ______________

16 *How to Dissect and Solve Word Problems: Multiplication and Division*

Solve the following problems:

1. Beverly plans on charging $16 per ticket for a buffet dinner and dance party she is organizing. She is certain she can sell 70 tickets. How much money will she receive from the ticket sales?

   $1,120

2. The total number of pages in a 23-volume set of encyclopedias is 10,327. What is the number of pages per volume if there are an equal number in each?

   449 pages

UNIT Name ______________________ Date ______________

16 ***How to Dissect and Solve Word Problems: Multiplication and Division***

3. What is the cost of 32 yards of fabric if it costs $18 per yard?

   $576

4. Mark has set a goal of saving a total of $2,482. He calculates that it will take 17 months to reach his goal. How much money must he save each month?

   $146

5. Clarisse records the attendance figures at a ballpark to be 38,716 people on each of 13 days. What is the total attendance?

   503,308 people

6. The product of two numbers is 5,301. If one number is 57 what is the other number?

   93

7. Dr. Gough's office booked an equal number of appointments each day for 1,081 patients over 23 days. How many appointments were booked each day?

   47 apppointments

8. Northern Bank and Trust is offering six condominiums for sale at $86,499 each. What is the total sales figure for the six condos?

   $518,994

9. Margaret spends $13 per week to park her car. She does so for 48 weeks each year. What is her total yearly expense?

   $624

10. Karen knows the cost of a violin to be $478. How much money must she raise to supply 23 students with violins?

    $10,994

11. If 156 feet of decking cost $468, how much would 175 feet cost at the same rate?

$525

12. Ace agrees to complete a project for Manny in 16 days. If his labor cost is $96 per day, what is the total labor cost?

$1,536

13. Joe finds that his car's gas tank holds 16 gallons. He estimates that he traveled 416 on one tank of gas. How many miles per gallon did he drive?

26 miles

14. The total number of people visiting a theme park over 7 days was 67,571. At this rate, how many people would visit the park in 30 days?

289,590 people

15. Each of 3,756 full-time students enrolled at a local community college pays $350 in fees each semester. What is the total amount of money collected in fees each semester?

$1,314,600

Name ______________________ Date ______________

Test

Answers

1. Write 6,713,476 in words. 1. six million, seven hundred thirteen thousand, four hundred seventy-six

2. Write two hundred sixty-one thousand, three hundred ninety-two in numeral form. 2. 261,392

3. Round to the nearest thousand: 47,673 3. 48,000

4. Add:

$$\begin{array}{r} 137 \\ 62 \\ +\ 298 \\ \hline \end{array}$$

4. 497

5. Subtract:

$$\begin{array}{r} 616 \\ -\ 97 \\ \hline \end{array}$$

5. 519

6. Estimate the sum by rounding each number to its first digit and adding:

$$\begin{array}{r} 6{,}943 \\ 3{,}560 \\ 1{,}495 \\ 2{,}846 \\ +\ 3{,}147 \\ \hline \end{array}$$

6. 18,000

7. The concert promoter printed 5,500 tickets for a local concert. If 1,173, 991, and 635 tickets were sold on three consecutive days, how many tickets remain? 7. 2,701 tickets

(Test continues on next page)

CHAPTER 1

Test *(Concluded)*

Answers

8. Multiply: $716 \times 59 =$

8. 42,244

9. Multiply:

a. $69{,}300 \times 480 =$

b. $69{,}300 \times 4{,}800 =$

c. $69{,}300 \times 48{,}000 =$

9a. 33,264,000

9b. 332,640,000

9c. 3,326,400,000

10. Divide: $35{,}750 \div 26 =$

10. 1,375

11. Divide: $48{,}873 \div 300 =$

11. 162 R273

12. A real estate office recently purchased three microcomputer systems for a total of $9,825. How much would it cost this office to purchase five additional systems if they are charged the same price per computer system?

12. $16,375

CHAPTER

2 Fractions

UNITS

UNIT 17 Identifying Fractions

Numbers that represent parts of a whole are called **fractions.** Fractions express the division of two numbers. For example, 3 divided by 5 is the fraction $\frac{3}{5}$ (read as "three-fifths"). The number below the line is called the **denominator,** and it expresses the number of equal parts into which the whole is divided (See Figure 2.17a). The **numerator** is the number above the line and it expresses how many of the equal parts are taken (see Figure 2.17b).

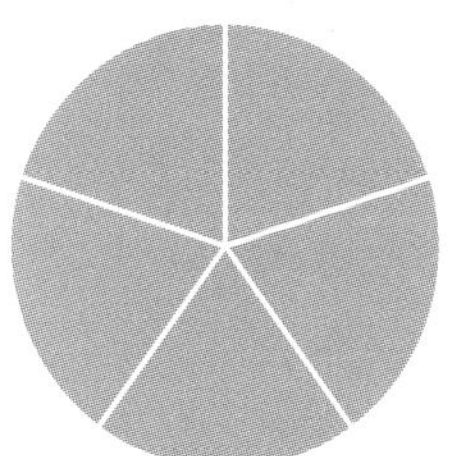

Figure 2.17a

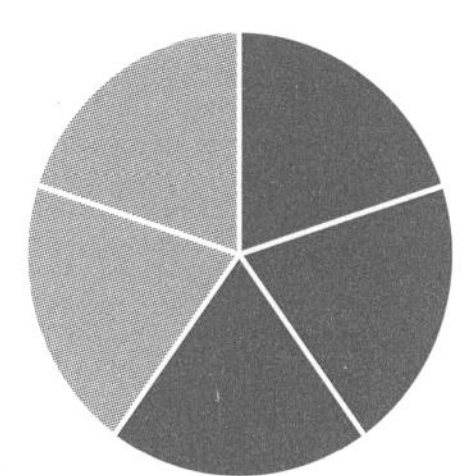

Figure 2.17b

Three types of fractions are:

1. **Proper Fractions**—the numerator is smaller than the denominator; for instance, $\frac{3}{5}$ or $\frac{1}{12}$ (read as "three-fifths or one-twelfth"). Proper fractions have values less than 1.
2. **Improper Fractions**—the numerator is greater than or equal to the denominator; for instance, $\frac{7}{4}$ or $\frac{2}{2}$ (read as "seven-fourths" and "two-halves"). Improper fractions have values greater than or equal to 1.
3. **Mixed Numbers**—a number expressing the sum of a whole number and a proper fraction, for instance $6\frac{1}{8}$ (read as "six and one-eighth").

EXAMPLE 1 **Identify the proper fractions, the improper fractions, and the mixed numbers in the following group of fractions.**

$$\frac{23}{4}, \frac{7}{8}, 4\frac{3}{7}, \frac{1}{9}, 22\frac{4}{9}, \frac{13}{15}, \frac{6}{5}, \frac{4}{4}$$

Answer: Proper fractions: $\frac{7}{8}, \frac{1}{9}, \frac{13}{15}$

Improper fractions: $\frac{23}{4}, \frac{6}{5}, \frac{4}{4}$

Mixed numbers: $4\frac{3}{7}, 22\frac{4}{9}$

PRACTICE 1 **Identify the proper fractions, the improper fractions, and the mixed numbers in the following group of fractions.**

$$1\frac{4}{9}, \frac{3}{4}, \frac{6}{5}, \frac{14}{16}, 3\frac{9}{12}, \frac{81}{3}, \frac{6}{7}, \frac{8}{8}$$

Answers: Proper fractions: $\frac{3}{4}$ $\frac{14}{16}$ $\frac{6}{7}$

Improper fractions: $\frac{6}{5}$ $\frac{81}{3}$ $\frac{8}{8}$

Mixed numbers: $1\frac{4}{9}$ $3\frac{9}{12}$

EXAMPLE 2 **Identify the fractions that represent 1.**

$\frac{3}{6}, \frac{4}{4}, \frac{144}{144}, \frac{16}{14}, \frac{22}{3}, 4\frac{4}{9}, \frac{15}{15}$ **Answers:** $\frac{4}{4}, \frac{144}{144}, \frac{15}{15}$

PRACTICE 2 **Identify the fractions that represent 1.**

$\frac{17}{15}, \frac{11}{12}, \frac{6}{6}, \frac{49}{49}, \frac{2}{7}, \frac{166}{166}, 1\frac{3}{13}$ **Answers:** $\frac{6}{6}$ $\frac{49}{49}$ $\frac{166}{166}$

EXAMPLE 3 **Write the following in fraction form.**

Answer: twelve-ninths $\frac{12}{9}$ four and six-fourteenths $4\frac{6}{14}$ three-elevenths $\frac{3}{11}$

PRACTICE 3 **Write the following in fraction form.**

nineteen-twentieths eight and three-seventeenths six-fourths

Answer: $\frac{19}{20}$ $8\frac{3}{17}$ $\frac{6}{4}$

UNIT **17** Name ________________ Date ________________

Identifying Fractions

Identify the following as proper (P) or improper (I) fractions or a mixed number (M). Indicate which fractions equal 1 by stating (1):

1. $\frac{7}{8}$ P
2. $13\frac{3}{9}$ M
3. $\frac{65}{65}$ 1
4. $\frac{2}{22}$ P
5. $\frac{14}{13}$ I
6. $144\frac{6}{7}$ M
7. $13\frac{3}{8}$ M
8. $\frac{16}{13}$ I
9. $\frac{12}{9}$ I
10. $\frac{100}{101}$ P
11. $\frac{195}{6}$ I
12. $\frac{19}{62}$ P
13. $\frac{5}{5}$ 1
14. $\frac{12}{8}$ I
15. $\frac{14}{17}$ P
16. $\frac{147}{147}$ 1
17. $\frac{19}{108}$ P
18. $\frac{17}{9}$ I
19. $\frac{22}{22}$ 1
20. $\frac{28}{31}$ P

UNIT 17 Name ______________________ Date ____________

Identifying Fractions

Express in fraction form:

21. six-ninths $\frac{6}{9}$
22. two-fifths $\frac{2}{5}$
23. seventeen-twenty-fifths $\frac{17}{25}$
24. two and two-sevenths $2\frac{2}{7}$
25. sixteen-thirds $\frac{16}{3}$
26. one and five-ninths $1\frac{5}{9}$
27. sixteen-thousandths $\frac{16}{1,000}$
28. fifteen-fifteenths $\frac{15}{15}$
29. ninety-nine-hundredths $\frac{99}{100}$
30. ten and one-fourth $10\frac{1}{4}$

UNIT

18 Conversion of Fractions: Improper Fractions and Mixed Numbers

In this unit, procedures for converting improper fractions to mixed numbers and for converting mixed numbers to improper fractions are presented.

To change an improper fraction to a mixed number:

1. Divide the numerator by the denominator.
2. Place the remainder over the denominator and add the quotient to this proper fraction.
3. If there is no remainder when dividing this numerator by the denominator, the answer is a whole number.

To change a mixed number to an improper fraction:

1. Multiply the denominator and the whole number.
2. Add the numerator to the product.
3. Write the sum over the denominator.

EXAMPLE 1 **Change the following to whole or mixed numbers.**

$\frac{17}{8} =$ **Answer:** $17 \div 8 = 2\,R1 \Rightarrow 2 + \frac{1}{8} = 2\frac{1}{8}$

$\frac{42}{6} =$ **Answer:** $42 \div 6 = 7$

PRACTICE 1 **Change the following to whole or mixed numbers.**

Answers: $\frac{23}{4}$ $5\frac{3}{4}$ $\frac{54}{6}$ 9

EXAMPLE 2 **Change $4\frac{2}{3}$ to an improper fraction.**

$4 \times 3 = 12$

$12 + 2 = 14$

Answer: $\frac{14}{3}$

PRACTICE 2 **Change $7\frac{3}{4}$ to an improper fraction.**

Answer: $\frac{31}{4}$

UNIT **18** Name ______ Date ______

Conversion of Fractions: Improper Fractions and Mixed Numbers

Change the following improper fractions to mixed or whole numbers:

1. $\frac{24}{9} =$ $2\frac{6}{9}$
2. $\frac{35}{5} =$ 7
3. $\frac{17}{7} =$ $2\frac{3}{7}$
4. $\frac{53}{4} =$ $13\frac{1}{4}$
5. $\frac{29}{6} =$ $4\frac{5}{6}$
6. $\frac{8}{3} =$ $2\frac{2}{3}$
7. $\frac{11}{4} =$ $2\frac{3}{4}$
8. $\frac{83}{10} =$ $8\frac{3}{10}$
9. $\frac{18}{5} =$ $3\frac{3}{5}$
10. $\frac{17}{17} =$ 1
11. $\frac{24}{6} =$ 4
12. $\frac{47}{6} =$ $7\frac{5}{6}$
13. $\frac{136}{136} =$ 1
14. $\frac{13}{8} =$ $1\frac{5}{8}$
15. $\frac{62}{5} =$ $12\frac{2}{5}$
16. $\frac{35}{3} =$ $11\frac{2}{3}$
17. $\frac{27}{2} =$ $13\frac{1}{2}$
18. $\frac{56}{8} =$ 7
19. $\frac{11}{10} =$ $1\frac{1}{10}$
20. $\frac{47}{4} =$ $11\frac{3}{4}$

Change the following to improper fractions:

21. $4\frac{2}{3} =$ $\frac{14}{3}$
22. $1\frac{7}{8} =$ $\frac{15}{8}$
23. $11\frac{3}{9} =$ $\frac{102}{9}$
24. $1\frac{3}{17} =$ $\frac{20}{3}$
25. $6\frac{2}{7} =$ $\frac{44}{7}$
26. $2\frac{3}{5} =$ $\frac{13}{5}$
27. $12\frac{2}{3} =$ $\frac{38}{3}$
28. $6\frac{3}{7} =$ $\frac{45}{7}$
29. $11\frac{1}{2} =$ $\frac{23}{2}$
30. $5\frac{12}{20} =$ $\frac{112}{20}$
31. $8\frac{5}{9} =$ $\frac{77}{9}$
32. $6\frac{3}{4} =$ $\frac{27}{4}$
33. $7\frac{5}{9} =$ $\frac{68}{9}$
34. $5\frac{3}{5} =$ $\frac{28}{5}$
35. $6\frac{2}{5} =$ $\frac{32}{5}$
36. $14\frac{8}{20} =$ $\frac{288}{20}$
37. $21\frac{1}{2} =$ $\frac{43}{2}$
38. $6\frac{3}{9} =$ $\frac{57}{9}$
39. $1\frac{121}{140} =$ $\frac{261}{140}$
40. $12\frac{5}{6} =$ $\frac{77}{6}$

UNIT 19 Reducing Fractions

Reducing a fraction is the process of changing its form without changing its value by dividing the numerator and denominator by the largest possible number that divides into each evenly. It is always understood that a fraction should be reduced to lowest terms. When no other number (except 1) can divide both parts of the fraction evenly, it is reduced to lowest terms.

To reduce a fraction to lowest terms:

1. Divide the numerator and denominator by the largest possible divisor.
2. Repeat the process, if necessary, until the fraction is reduced to lowest terms.
3. Change any improper fractions to mixed numbers after reducing.
4. Reduce a mixed number by reducing the fraction part.

The following rules should be helpful in finding the divisors of a number.

- **a** If its last digit ends in 0, 2, 4, 6, or 8, divide the number by 2.
- **b** If the sum of its digits is divisible by 3, divide the number by 3.
- **c** If its last two digits is divisible by 4, divide the number by 4.
- **d** If its last digit is 0 or 5, divide the number by 5.
- **e** If the number is even *and* 3 divides the sum of its digits, divide the number by 6.
- **f** If the sum of its digits is divisible by 9, divide the number by 9.
- **g** If the last digit is 0, divide the number by 10.

EXAMPLE 1 **Reduce $\frac{18}{96}$ to lowest terms:**

Answer: $\frac{18}{96} = \frac{18 \div 6}{96 \div 6} = \frac{3}{16}$ or $\frac{18}{96} = \frac{18 \div 2}{96 \div 2} = \frac{9}{48} = \frac{9 \div 3}{48 \div 3} = \frac{3}{16}$

PRACTICE 1 **Reduce $\frac{27}{54}$ to lowest terms:**

Answer: $\frac{1}{2}$

EXAMPLE 2 **Reduce $6\frac{25}{35}$ to lowest terms:**

Answer: $6\frac{25}{35} = 6\frac{25 \div 5}{35 \div 5} = 6\frac{5}{7}$

PRACTICE 2 **Reduce $4\frac{12}{27}$ to lowest terms:**

Answer: $4\frac{4}{9}$

UNIT 19 Name ______________________ Date ______________

Reducing Fractions

Reduce to lowest terms:

1. $\frac{10}{12} = \frac{5}{6}$
2. $\frac{36}{64} = \frac{9}{16}$
3. $\frac{19}{38} = \frac{1}{2}$
4. $\frac{25}{100} = \frac{1}{4}$
5. $\frac{6}{36} = \frac{1}{6}$
6. $\frac{14}{28} = \frac{1}{2}$
7. $\frac{33}{165} = \frac{1}{5}$
8. $\frac{9}{36} = \frac{1}{4}$
9. $\frac{25}{45} = \frac{5}{9}$
10. $\frac{26}{34} = \frac{13}{17}$
11. $\frac{91}{154} = \frac{13}{22}$
12. $\frac{45}{63} = \frac{5}{7}$
13. $\frac{15}{20} = \frac{3}{4}$
14. $\frac{144}{332} = \frac{36}{83}$
15. $\frac{12}{36} = \frac{1}{3}$
16. $\frac{80}{45} = 1\frac{7}{9}$
17. $\frac{16}{12} = 1\frac{1}{3}$
18. $\frac{36}{16} = 2\frac{1}{4}$
19. $\frac{66}{30} = 2\frac{1}{5}$
20. $\frac{52}{16} = 3\frac{1}{4}$
21. $\frac{38}{8} = 4\frac{3}{4}$
22. $\frac{64}{72} = \frac{8}{9}$
23. $\frac{98}{102} = \frac{49}{51}$
24. $\frac{20}{30} = \frac{2}{3}$
25. $\frac{65}{75} = \frac{13}{15}$
26. $\frac{22}{33} = \frac{2}{3}$
27. $4\frac{3}{9} = 4\frac{1}{3}$
28. $18\frac{24}{27} = 18\frac{8}{9}$
29. $5\frac{63}{90} = 5\frac{7}{10}$
30. $15\frac{20}{48} = 15\frac{5}{12}$
31. $22\frac{21}{36} = 22\frac{7}{12}$
32. $10\frac{14}{24} = 10\frac{7}{12}$
33. $7\frac{16}{24} = 7\frac{2}{3}$

Business Applications:

34. Nine out of every 72 electric razors manufactured by a small electronics firm are defective. Express this quantity as a reduced fraction.

$\frac{1}{8}$

35. Nichols Wholesale Glassware knows that out of 350 pieces of crystal it shipped to a retailer, 30 arrived broken. Express this relationship as a reduced fraction.

$\frac{3}{35}$

UNIT

20 Raising Fractions to Higher Terms

Raising a fraction to higher terms is the process of creating an equal fraction with a larger denominator. This is accomplished by multiplying the numerator and denominator of the fraction by the same nonzero number. The new fraction created will be equal to the old fraction, because we have only multiplied the old fraction by 1. The steps for raising a fraction to higher terms when the denominator is known are:

1. Divide the old denominator into the desired new, larger denominator.
2. Multiply the numerator by the quotient and place it above the new denominator.

EXAMPLE 1 **Raise each fraction to higher terms.**

a. $\frac{3}{7} = \frac{?}{42}$ $42 \div 7 = 6$, $6 \times 3 = 18$

Answers: $\frac{3}{7} = \frac{18}{42}$

b. $\frac{16}{9} = \frac{?}{27}$ $27 \div 9 = 3$, $3 \times 16 = 48$

$\frac{16}{9} = \frac{48}{27}$

PRACTICE 1 **Raise each fraction to higher terms.**

a. $\frac{4}{7} = \frac{?}{28}$ **Answer:** 16

b. $\frac{15}{6} = \frac{?}{24}$ **Answer:** 60

UNIT 20 Name ________ Date ________

Raising Fractions to Higher Terms

Raise each fraction to higher terms:

1. $\frac{5}{9} = \frac{?}{45}$ 25
2. $\frac{7}{8} = \frac{?}{40}$ 35
3. $\frac{5}{2} = \frac{?}{24}$ 60
4. $\frac{9}{16} = \frac{?}{48}$ 27
5. $\frac{5}{6} = \frac{?}{30}$ 25
6. $\frac{7}{6} = \frac{?}{42}$ 49
7. $\frac{2}{8} = \frac{?}{16}$ 4
8. $\frac{9}{12} = \frac{?}{72}$ 54

9. $\frac{4}{6} = \frac{?}{54}$ 36 10. $\frac{15}{14} = \frac{?}{56}$ 60 11. $\frac{1}{12} = \frac{?}{144}$ 12 12. $\frac{4}{10} = \frac{?}{1,000}$ 400

13. $\frac{2}{9} = \frac{?}{27}$ 6 14. $\frac{1}{8} = \frac{?}{64}$ 8 15. $\frac{7}{3} = \frac{?}{9}$ 21 16. $\frac{18}{27} = \frac{?}{135}$ 90

17. $\frac{5}{8} = \frac{?}{160}$ 100 18. $\frac{1}{9} = \frac{?}{180}$ 20 19. $\frac{6}{11} = \frac{?}{121}$ 66 20. $\frac{8}{5} = \frac{?}{100}$ 160

21. $\frac{4}{3} = \frac{?}{18}$ 24 22. $\frac{14}{12} = \frac{?}{84}$ 98 23. $\frac{7}{3} = \frac{?}{81}$ 189 24. $\frac{3}{4} = \frac{?}{64}$ 48

25. $\frac{9}{6} = \frac{?}{18}$ 27 26. $\frac{4}{1} = \frac{?}{12}$ 48 27. $\frac{1}{5} = \frac{?}{100}$ 20 28. $\frac{7}{8} = \frac{?}{56}$ 49

Business Applications:

29. Mega Byte Technology knows that 1 out of every 7 computer monitors it ships arrives damaged. At this rate, how many damaged monitors can it anticipate out of a shipment of 210?

30 monitors

30. Out of every 100 season ticket holders to a minor league baseball team, 72 are renewed for the upcoming season. Express this relationship as an equivalent fraction with a denominator of 3,500.

$\frac{2,520}{3,500}$

UNIT

21 Adding Fractions with Like Denominators

The addition of fractions with like denominators is a simple procedure.

1. Add the numerators of the fractions.
2. Place the total over the common denominator.
3. If mixed numbers are being added, add the whole number parts.
4. Reduce the fraction to lowest terms or convert any improper fractions to mixed numbers.

EXAMPLE 1 **Add the following.**

a. $\frac{4}{9} + \frac{3}{9}$

b. $2\frac{1}{8} + 5\frac{3}{8}$

c. $\frac{2}{7} + \frac{4}{7} + \frac{5}{7} + \frac{6}{7}$

Answers:

a. $\frac{4}{9} + \frac{3}{9} = \frac{7}{9}$

b. $2\frac{1}{8} + 5\frac{3}{8} = 7\frac{4}{8} = 7\frac{1}{2}$

c. $\frac{2}{7} + \frac{4}{7} + \frac{5}{7} + \frac{6}{7} = \frac{17}{7} = 2\frac{3}{7}$

PRACTICE 1 **Add the following.**

a. $\frac{4}{11} + \frac{5}{11} =$ **Answer:** $\frac{9}{11}$

b. $3\frac{2}{7} + 5\frac{1}{7} =$ **Answer:** $8\frac{3}{7}$

c. $\frac{7}{8} + \frac{3}{8} + \frac{1}{8} + \frac{5}{8} + \frac{1}{8} =$ **Answer:** $2\frac{1}{8}$

UNIT 21 Name ____________ Date ____________

Adding Fractions with Like Denominators

Add the following (reduce to lowest terms if necessary):

1. $\frac{1}{7} + \frac{3}{7} =$ $\frac{4}{7}$
2. $\frac{1}{2} + \frac{5}{2} =$ 3
3. $\frac{3}{12} + \frac{5}{12} =$ $\frac{2}{3}$
4. $\frac{2}{8} + \frac{4}{8} =$ $\frac{3}{4}$

5. $\frac{2}{7} + \frac{5}{7} =$ 1

6. $\frac{12}{27} + \frac{8}{27} =$ $\frac{20}{27}$

7. $\frac{3}{9} + \frac{4}{9} =$ $\frac{7}{9}$

8. $\frac{2}{7} + \frac{1}{7} + \frac{5}{7} =$ $1\frac{1}{7}$

9. $\frac{25}{37} + \frac{2}{37} + \frac{8}{37} =$ $\frac{35}{37}$

10. $\frac{7}{16} + \frac{11}{16} + \frac{1}{16} =$ $1\frac{3}{16}$

11. $3\frac{2}{8} + 5\frac{1}{8} =$ $8\frac{3}{8}$

12. $6\frac{2}{5} + 7\frac{1}{5} =$ $13\frac{3}{5}$

13. $9\frac{1}{16} + 7\frac{5}{16} =$ $16\frac{3}{8}$

14. $4\frac{3}{19} + 7\frac{3}{19} =$ $11\frac{6}{19}$

15. $7\frac{5}{14} + 3\frac{7}{14} =$ $10\frac{6}{7}$

16. $4\frac{8}{9} + 3\frac{4}{9} =$ $8\frac{1}{3}$

17. $5\frac{5}{6} + 6\frac{1}{6} =$ 12

18. $\frac{7}{12} + \frac{13}{12} =$ $1\frac{2}{3}$

19. $\frac{8}{9} + \frac{7}{9} + \frac{3}{9} =$ 2

20. $3\frac{1}{5} + 4\frac{3}{5} + 6\frac{2}{5} =$ $14\frac{1}{5}$

21. $7\frac{10}{18} + 12\frac{6}{18} + 3\frac{5}{18} =$ $23\frac{1}{6}$

22. $8\frac{7}{13} + 7\frac{12}{13} + 1\frac{7}{13} =$ 18

23. $\frac{7}{25} + \frac{4}{25} + \frac{13}{25} + \frac{9}{25} =$ $1\frac{8}{25}$

Business Applications:

24. A certain stock closed at $24\frac{1}{8}$ the previous day. Its gain for the present day is $\frac{3}{8}$. Add these quantities to determine its current price.

$24\frac{1}{2}$

25. A stock closed at $61\frac{5}{8}$ yesterday. It has gained $\frac{3}{8}$ since yesterday. Add these quantities to find its current price.

62

UNIT

22 Lowest Common Denominator

Because fractions with like denominators only can be added and subtracted, we must sometimes change the denominator of the fractions we wish to add or subtract to the same number. This number is called the **lowest common denominator (LCD).** The LCD is the smallest possible number into which all the denominators in the problem can be divided. The LCD can be found by trial and error or by using prime numbers. A **prime number** is a number divisible only by itself and the number 1. For example, 3 is a prime number, 6 is not (6 is divisible not only by itself and 1 but also by 2 or 3). A partial list of whole prime numbers is 2, 3, 5, 7, 11, 13, 17, 19, 23, 31, 37, 41, 43.

Any number can be expressed as the product of prime numbers. We accomplish this by following these steps:

1. Divide the number by the first prime number that can possibly divide it evenly.
2. Continue dividing each quotient by the first possible prime number until arriving at the quotient 1.
3. The original number can now be expressed as the product of the prime divisors.

When determining an LCD for a group of fractions using prime numbers:

1. Express the denominator of each fraction as the product of primes.
2. List each different prime number from these expressions.
3. Each prime from the list must now be expressed in the LCD the greatest number of times it appears in any single denominator.
4. The LCD is formed by multiplying these primes together.

EXAMPLE 1 **Express each as the product of prime numbers.**

a. 48

$48 \div 2 = 24$
$24 \div 2 = 12$
$12 \div 2 = 6$
$6 \div 2 = 3$
$3 \div 3 = 1$

b. 630

$630 \div 2 = 315$
$315 \div 3 = 105$
$105 \div 3 = 35$
$35 \div 5 = 7$
$7 \div 7 = 1$

Answers: a. $48 = 2 \times 2 \times 2 \times 2 \times 3$ b. $630 = 2 \times 3 \times 3 \times 5 \times 7$

PRACTICE 1 **Express each as the product of prime numbers.**

Answers: a. $72 =$ $2 \times 2 \times 2 \times 3 \times 3$ b. $360 =$ $2 \times 2 \times 2 \times 3 \times 3 \times 5$

EXAMPLE 2 **Find the LCD for the following set of fractions.**

$\frac{1}{4}, \frac{1}{8}, \frac{1}{9}, \frac{1}{12}$

Answer: $\frac{1}{\boxed{2 \cdot 2}}, \frac{1}{\boxed{2 \cdot 2 \cdot 2}}, \frac{1}{\boxed{3 \cdot 3}}, \frac{1}{\boxed{2 \cdot 2 \cdot 3}}$

$\boxed{2}$, $\boxed{3}$ } List of different prime numbers

1. Express each denominator as the product of primes.
2. List each different prime number contained in the denominators.
3. The greatest number of times 2 appears in a single denominator is *three times.* It appears in the LCD three times. The greatest number of times 3 appears in a single denominator is *twice.* It appears in the LCD twice.
4. LCD = $\boxed{2 \cdot 2 \cdot 2} \cdot \boxed{3 \cdot 3} = 72$

PRACTICE 2 **Find the LCD for the following set of fractions.**

$\frac{1}{10}, \frac{1}{12}, \frac{3}{4}, \frac{7}{8} =$ **Answer:** 120

UNIT **22** Name ______________________ Date ____________

Lowest Common Denominator

Express each of the following as the product of primes:

1. 64 = $2 \times 2 \times 2 \times 2 \times 2 \times 2$
2. 21 = 3×7
3. 27 = $3 \times 3 \times 3$
4. 45 = $3 \times 3 \times 5$
5. 40 = $2 \times 2 \times 2 \times 5$
6. 75 = $3 \times 5 \times 5$
7. 100 = $2 \times 2 \times 5 \times 5$
8. 96 = $2 \times 2 \times 2 \times 2 \times 2 \times 3$
9. 120 = $2 \times 2 \times 2 \times 3 \times 5$
10. 240 = $2 \times 2 \times 2 \times 2 \times 3 \times 5$

Find the LCD for each of the following groups of fractions:

11. $\frac{1}{6}, \frac{7}{20} =$ 60
12. $\frac{13}{24}, \frac{8}{12} =$ 24
13. $\frac{1}{27}, \frac{3}{15} =$ 135
14. $\frac{1}{15}, \frac{2}{35}, \frac{1}{9} =$ 315
15. $\frac{1}{6}, \frac{3}{9}, \frac{4}{12}, \frac{2}{8} =$ 72
16. $\frac{5}{32}, \frac{1}{12}, \frac{7}{48}, \frac{1}{64} =$ 192
17. $\frac{11}{16}, \frac{2}{32}, \frac{5}{8}, \frac{4}{12} =$ 96
18. $\frac{3}{9}, \frac{1}{24}, \frac{5}{16}, \frac{10}{30} =$ 720
19. $\frac{13}{18}, \frac{17}{32}, \frac{5}{64}, \frac{3}{25} =$ 14,400
20. $\frac{5}{24}, \frac{3}{32}, \frac{1}{12}, \frac{3}{10} =$ 480

UNIT

23 Adding Fractions with Unlike Denominators

As stated before, to add fractions the denominators must all be the same. The LCD can be found by inspection or by using the prime number approach described in Unit 22.

The general steps for adding fractions with unlike denominators are:

1. Find the LCD.
2. Change each fraction to an equal fraction by dividing the LCD by the old denominator and by multiplying the old numerator by the quotient.
3. Add the numerators (and whole numbers if necessary) and place the result over the LCD.
4. Reduce to lowest terms or change any improper fractions to mixed numbers.

EXAMPLE 1 **Add: $4\frac{3}{4} + 2\frac{1}{8} + 6\frac{1}{3}$.**

$$4\frac{3}{4} = 4\frac{18}{24}$$

1. The LCD is 24.

$$2\frac{1}{8} = 2\frac{3}{24}$$

2. Change each numerator.

$$+\ 6\frac{1}{3} = 6\frac{8}{24}$$

3. Add the new fractions.

$$12\frac{29}{24}$$

$$\frac{29}{24} = 1\frac{5}{24}$$

4. Change $\frac{29}{24}$ to a mixed number.
5. Add 12 + 1.

Answer: $12 + 1\frac{5}{24} = 13\frac{5}{24}$

6. Set 13 next to $\frac{5}{24}$.

PRACTICE 1 **Add: $3\frac{1}{5} + 5\frac{2}{4} + 1\frac{2}{6}$.**

Answer: $10\frac{1}{30}$

UNIT 23 Name ______________________ Date ____________

Adding Fractions with Unlike Denominators

Add and reduce to lowest terms:

1. $\frac{1}{2} + \frac{1}{4}$ — $\frac{3}{4}$
2. $\frac{3}{4} + \frac{5}{8}$ — $1\frac{3}{8}$
3. $\frac{5}{4} + \frac{1}{12}$ — $1\frac{1}{3}$
4. $\frac{1}{2} + \frac{3}{4}$ — $1\frac{1}{4}$
5. $\frac{1}{4} + \frac{1}{3}$ — $\frac{7}{12}$
6. $\frac{5}{6} + \frac{2}{9} + \frac{1}{3}$ — $1\frac{7}{18}$
7. $\frac{3}{4} + \frac{3}{8} + \frac{2}{12}$ — $1\frac{7}{24}$
8. $\frac{7}{8} + \frac{2}{3} + \frac{4}{6}$ — $2\frac{5}{24}$
9. $\frac{2}{3} + \frac{5}{9} + \frac{1}{10}$ — $1\frac{29}{90}$
10. $\frac{3}{5} + \frac{1}{10} + \frac{6}{7}$ — $1\frac{39}{70}$
11. $8\frac{1}{4} + 12\frac{1}{2} + 9\frac{1}{3}$ — $30\frac{1}{12}$
12. $5\frac{1}{3} + 9\frac{1}{4} + 6\frac{1}{2}$ — $21\frac{1}{12}$
13. $12\frac{1}{5} + 7\frac{1}{2} + 9\frac{1}{4}$ — $28\frac{19}{20}$
14. $6\frac{1}{4} + 4\frac{1}{3} + 5\frac{1}{5}$ — $15\frac{47}{60}$
15. $4\frac{1}{4} + 7\frac{2}{5} + 8\frac{1}{8}$ — $19\frac{31}{40}$
16. $12\frac{1}{2} + 5\frac{5}{6} + 12\frac{1}{3}$ — $30\frac{2}{3}$
17. $4\frac{1}{4} + 4\frac{1}{3} + 19\frac{1}{2}$ — $28\frac{1}{12}$
18. $3\frac{5}{16} + 6\frac{1}{8} + 13\frac{3}{4}$ — $23\frac{3}{16}$
19. $10\frac{1}{3} + 11\frac{5}{6} + 4\frac{4}{9}$ — $26\frac{11}{18}$
20. $24\frac{5}{8} + 5\frac{2}{3} + 11\frac{1}{4}$ — $41\frac{13}{24}$

Business Applications:

21. PGR stock gained $\frac{3}{8}$ over yesterday's closing price of $23\frac{1}{4}$. Add these quantities to find this stock's final price.

    $23\frac{5}{8}$

22. In unpacking a shipment of canned soup, a retailer finds $\frac{1}{5}$ of the cans in one case to be damaged and $\frac{3}{8}$ of the cans in another to be crushed. Combine these fractions to express the total portion of the shipment that is damaged.

    $\frac{23}{40}$ of the shipment

UNIT

24 Subtracting Fractions with Like Denominators

Subtracting fractions with like denominators is completed in the same way as addition.

1 Subtract the numerators.

2 Place the difference over the common denominator.

3 If mixed numbers are being subtracted, subtract the whole numbers.

4 Reduce to lowest terms or change any improper fractions to mixed numbers.

EXAMPLE 1 **Subtract: $4\frac{9}{16} - 2\frac{7}{16}$.**

$$\begin{array}{r} 4\frac{9}{16} \\ -\ 2\frac{7}{16} \\ \hline 2\frac{2}{16} = 2\frac{1}{8} \end{array}$$

1. Subtract the numerators.
2. Subtract whole numbers.
3. Reduce to lowest terms.

Answer: $2\frac{2}{16} = 2\frac{1}{8}$

PRACTICE 1 **Subtract: $7\frac{13}{24} - 4\frac{5}{24}$.**

Answer: $3\frac{1}{3}$

UNIT 24 Name ______________________ Date ______________

Subtracting Fractions with Like Denominators

Subtract the following:

1. $\frac{4}{5} - \frac{1}{5} =$ $\frac{3}{5}$

2. $\frac{13}{17} - \frac{5}{17} =$ $\frac{8}{17}$

3. $\frac{14}{15} - \frac{5}{15} =$ $\frac{3}{5}$

4. $\frac{6}{9} - \frac{3}{9} =$ $\frac{1}{3}$

5. $\frac{18}{20} - \frac{4}{20} =$ $\frac{7}{10}$

6. $\frac{9}{10} - \frac{7}{10} =$ $\frac{1}{5}$

7. $\frac{19}{36} - \frac{17}{36} =$ $\frac{1}{18}$

8. $\frac{5}{6} - \frac{1}{6} =$ $\frac{2}{3}$

9. $\frac{7}{8} - \frac{5}{8} =$ $\frac{1}{4}$

10. $\frac{11}{12} - \frac{3}{12} =$ $\frac{2}{3}$

11. $\frac{7}{12} - \frac{2}{12} =$ $\frac{5}{12}$

12. $\frac{7}{15} - \frac{3}{15} =$ $\frac{4}{15}$

13. $\frac{5}{6} - \frac{3}{6} =$ $\frac{1}{3}$

14. $\frac{12}{22} - \frac{6}{22} =$ $\frac{3}{11}$

15. $\frac{14}{56} - \frac{7}{56} =$ $\frac{1}{8}$

16. $15\frac{7}{8} - 6\frac{5}{8}$ = $9\frac{1}{4}$

17. $14\frac{9}{10} - 8\frac{7}{10}$ = $6\frac{1}{5}$

18. $11\frac{19}{24} - 5\frac{5}{24}$ = $6\frac{7}{12}$

19. $9\frac{5}{6} - 6\frac{1}{6}$ = $3\frac{2}{3}$

20. $9\frac{11}{12} - 3\frac{2}{12}$ = $6\frac{3}{4}$

21. $10\frac{7}{15} - 5\frac{2}{15}$ = $5\frac{1}{3}$

22. $28\frac{2}{3} - 14\frac{1}{3}$ = $14\frac{1}{3}$

23. $20\frac{3}{4} - 19\frac{1}{4}$ = $1\frac{1}{2}$

24. $12\frac{7}{12} - 4\frac{6}{12}$ = $8\frac{1}{12}$

25. $6\frac{3}{8} - 4\frac{1}{8}$ = $2\frac{1}{4}$

26. $8\frac{7}{12} - 6\frac{2}{12}$ = $2\frac{5}{12}$

27. $14\frac{3}{50} - 12\frac{1}{50}$ = $2\frac{1}{25}$

28. $7\frac{15}{30} - 3\frac{6}{30}$ = $4\frac{3}{10}$

Business Applications:

29. A bookkeeper for a small accounting firm can process a file of tax returns in $7\frac{1}{4}$ hours. A second bookkeeper processes the same file of returns in $8\frac{3}{4}$ hours. Subtract these fractions to find the difference in the times each has worked.

$1\frac{1}{2}$ hours

30. A tarpaulin manufacturer finds two tarpaulins to be of different lengths when, in fact, they should be the same. One measures $22\frac{7}{8}$ yards and the other $19\frac{5}{8}$. Subtract these fractions to find the difference in their lengths.

$3\frac{1}{4}$ yards

UNIT

25 Subtracting Fractions with Unlike Denominators

When subtracting fractions with unlike denominators, follow these steps:

1. Rewrite any mixed numbers as improper fractions.
2. Rewrite any whole numbers as a fraction, placing the whole number above 1.
3. Find the LCD.
4. Change each fraction to an equivalent fraction using the LCD.
5. Subtract the numerator and place the difference over the LCD.
6. Reduce to lowest terms, rewriting any improper fractions as mixed numbers, if necessary.

EXAMPLE 1 **Subtract: $\frac{7}{8} - \frac{1}{3}$.**

$$\frac{7}{8} = \frac{21}{24}$$
$$-\ \frac{1}{3} = -\ \frac{8}{24}$$

Answer: $\frac{13}{24}$

1. The LCD = 24.
2. Change each fraction to an equivalent fraction.
3. Subtract and reduce to lowest terms.

PRACTICE 1 **Subtract: $\frac{8}{11} - \frac{3}{7}$.**

Answer: $\frac{23}{77}$

EXAMPLE 2 **Subtract: $4\frac{1}{3} - 1\frac{7}{8}$.**

$$4\frac{1}{3} = \frac{13}{3} = \frac{104}{24}$$
$$-\ 1\frac{7}{8} = -\ \frac{15}{8} = -\ \frac{45}{24}$$
$$\frac{59}{24}$$

Answer: $\frac{59}{24} = 2\frac{11}{24}$

1. Rewrite mixed numbers as improper fractions.
2. The LCD = 24.
3. Change each fraction to an equivalent fraction.
4. Subtract.
5. Rewrite as a mixed number.

PRACTICE 2 **Subtract: $7\frac{1}{5} - 4\frac{2}{3}$.**

Answer: $2\frac{8}{15}$

EXAMPLE 3 **Subtract: $11 - 5\frac{3}{4}$.**

$$\begin{array}{rcrcr} 11 & = & \frac{11}{1} & = & \frac{44}{4} \\ -\ 5\frac{3}{4} & = & -\ \frac{23}{4} & = & -\ \frac{23}{4} \\ \hline & & & & \frac{21}{4} \end{array}$$

$$\frac{21}{4} = 5\frac{1}{4}$$

1. Rewrite mixed numbers.
2. Rewrite whole number.
3. The LCD = 4.
4. Change each fraction to an equivalent fraction.
5. Subtract.
6. Rewrite as a mixed number.

PRACTICE 3 **Subtract: $14 - 3\frac{2}{3}$.**

Answer: $10\frac{1}{3}$

UNIT **25** Name ____________________ Date ____________

Subtracting Fractions with Unlike Denominators

Subtract the following:

1. $\frac{1}{2} - \frac{1}{4} =$ $\frac{1}{4}$
2. $\frac{3}{4} - \frac{1}{2} =$ $\frac{1}{4}$
3. $\frac{7}{8} - \frac{3}{4} =$ $\frac{1}{8}$
4. $\frac{3}{4} - \frac{1}{6} =$ $\frac{7}{12}$
5. $\frac{5}{9} - \frac{1}{3} =$ $\frac{2}{9}$
6. $\frac{8}{9} - \frac{5}{6} =$ $\frac{1}{18}$
7. $\frac{11}{12} - \frac{3}{4} =$ $\frac{1}{6}$
8. $\frac{3}{4} - \frac{7}{12} =$ $\frac{1}{6}$
9. $\frac{11}{16} - \frac{2}{3} =$ $\frac{1}{48}$
10. $\frac{8}{9} - \frac{4}{11} =$ $\frac{52}{99}$

UNIT **25** Name ______________________ Date ______________

Subtracting Fractions with Unlike Denominators

11. $4 - 1\frac{2}{3} =$ $2\frac{1}{3}$

12. $6 - 4\frac{3}{4} =$ $1\frac{1}{4}$

13. $5 - 2\frac{5}{8} =$ $2\frac{3}{8}$

14. $3 - 1\frac{1}{4} =$ $1\frac{3}{4}$

15. $9 - 5\frac{5}{6} =$ $3\frac{1}{6}$

16. $7\frac{1}{2} - 5\frac{7}{8} =$ $1\frac{5}{8}$

17. $6\frac{1}{2} - 4\frac{2}{3} =$ $1\frac{5}{6}$

18. $12\frac{1}{2} - 8\frac{7}{8} =$ $3\frac{5}{8}$

19. $13\frac{1}{4} - \frac{7}{8} =$ $12\frac{3}{8}$

20. $10\frac{2}{3} - 3\frac{3}{5} =$ $7\frac{1}{15}$

21. $14\frac{1}{2} - 5\frac{5}{8} =$ $8\frac{7}{8}$

22. $8 - 5\frac{5}{16} =$ $2\frac{11}{16}$

23. $7 - 5\frac{4}{5} =$ $1\frac{1}{5}$

24. $6 - 4\frac{1}{8} =$ $1\frac{7}{8}$

25. $5\frac{1}{3} - 2\frac{1}{9} =$ $3\frac{2}{9}$

Business Applications:

26. Times Mirror stock reached a high of $72\frac{1}{4}$ per share on Monday. At the end of the day, the stock dropped to $66\frac{3}{8}$ per share. Subtract to find how much the stock dropped from its high.

$5\frac{7}{8}$

27. A parcel delivery service estimates it will take $4\frac{1}{2}$ hours to deliver the cargo it has contracted to ship. Assume $2\frac{1}{4}$ hours has been spent making some of the deliveries. Subtract these fractions to find how much time remains to make the entire delivery.

$2\frac{1}{4}$ hours

UNIT

26 How to Dissect and Solve Word Problems: Adding and Subtracting Fractions

The chart that follows serves as a guide for dissecting and solving the word problems in this unit. Review, if necessary, the general suggestions for solving word problems in Unit 8 of Chapter 1.

Gather the facts.	What am I solving for?	What must I need to know or calculate before solving the problem?	Key points to remember.

EXAMPLE 1 **Two thirds of a cup of liquid is added to a mixture that already has $1\frac{1}{2}$ cups of liquid in it. What is the total amount of liquid?**

Gather the facts.	What am I solving for?	What must I need to know or calculate before solving the problem?	Key points to remember.
Original amount: $1\frac{1}{2}$ cups Added amount: $\frac{2}{3}$ cups	Total amount of liquid	Original amount + Added amount = Total amount	Common denominators are necessary when adding. Reduce answer if necessary.

$$
\begin{array}{rl}
1\frac{1}{2} = & 1\frac{3}{6} \\
+ \quad \frac{2}{3} = & \frac{4}{6} \\
\hline
& 1\frac{7}{6} = 1 + 1\frac{1}{6}
\end{array}
$$

Answer: $= 2\frac{1}{6}$ cups

PRACTICE 1 **Michael reads $\frac{1}{3}$ of a book on Monday and an additional $\frac{3}{8}$ of the book on Tuesday. How much of the book has he read?**

Answer: $\frac{17}{24}$ *of the book*

EXAMPLE 2 **Paul walked a total of $2\frac{1}{2}$ miles on Friday. If $\frac{2}{3}$ of a mile was uphill and $\frac{5}{8}$ of a mile was downhill, what portion of the route was level ground?**

Gather the facts.	What am I solving for?	What must I need to know or calculate before solving the problem?	Key points to remember.
Total = $2\frac{1}{2}$ miles Uphill = $\frac{2}{3}$ mile Downhill = $\frac{5}{8}$ mile	What part of the route is level?	Total mileage − (uphill + downhill mileage) = Level ground mileage	Common denominators are needed. Reduce the answer if necessary.

$$\frac{2}{3} = \frac{16}{24}$$
$$+\ \frac{5}{8} = \frac{15}{24}$$
$$\frac{31}{24}$$

$$2\frac{1}{2} = \frac{5}{2} = \frac{60}{24}$$
$$-\ \frac{31}{24} = -\ \frac{31}{24} = \frac{31}{24}$$

Answer: $\frac{29}{24} = 1\frac{5}{24}$ mile

PRACTICE 2 **A lot of $7\frac{3}{4}$ acres has been subdivided, and two lots of $1\frac{1}{2}$ acres and $1\frac{7}{8}$ acres have already been sold. How much remaining acreage is for sale?**

Answer: $4\frac{3}{8}$ *acres*

UNIT **26** Name ______________________ Date ______________

How to Dissect and Solve Word Problems: Adding and Subtracting Fractions

Solve the following problems:

1. If a bag of dried fruit contains $1\frac{1}{4}$ pounds of apricots and $1\frac{1}{2}$ pounds of pineapple, how many pounds does the bag weigh?

   $2\frac{3}{4}$ pounds

2. Mike spent $3\frac{2}{3}$ hours painting the ceiling and walls of a room at his rental property. If $1\frac{1}{2}$ hours was spent painting the ceiling, what amount of time was spent painting walls?

   $2\frac{1}{6}$ hours

3. On 3 days of the Jackson's family vacation it rained $\frac{3}{4}$ of an inch, $1\frac{1}{8}$ inches, and $2\frac{3}{16}$ inches. What was the total rainfall in inches?

   $4\frac{1}{16}$ inches

4. It takes $4\frac{1}{4}$ hours to cook a large turkey. Fred mistakenly tells his daughter that the turkey will be cooked in $2\frac{1}{2}$ hours. By how much time was Fred mistaken?

   $1\frac{3}{4}$ hours

5. In one workday, Diane spends $3\frac{1}{2}$ hours talking to patients, $\frac{3}{4}$ of an hour doing paperwork, $1\frac{1}{8}$ hours on the telephone, and $\frac{2}{3}$ of an hour doing other work. What is the total work time for the day?

   $6\frac{1}{24}$ hours

6. Tim is $73\frac{3}{8}$ inches tall, whereas Janice is $66\frac{4}{5}$ inches. How much taller is Tim than Janice?

   $6\frac{23}{40}$ inches

7. Maria ran $1\frac{1}{2}$ miles on Monday, $1\frac{3}{4}$ miles on Tuesday, $\frac{7}{8}$ of a mile on Wednesday, and 2 miles on Friday. What was the total mileage she ran?

   $6\frac{1}{8}$ miles

8. A curtain rod must be adjusted to accommodate a certain picture window. If the original rod of 6 feet must be shortened by $1\frac{3}{8}$ feet, what is the desired length of the rod?

   $4\frac{5}{8}$ feet

9. On 3 successive trips to the gas station, Art filled his truck's gas tank with $7\frac{7}{8}$ gallons, $3\frac{3}{4}$ gallons, and $9\frac{1}{3}$ gallons of gas. How many gallons of gas were put in altogether?

   $20\frac{23}{24}$ gallons

10. A stock's original selling price was $36\frac{1}{4}$ per share at the start of the day. By day's end it had dropped $\frac{7}{8}$ of a point. What was the selling price at day's end?

    $35\frac{3}{8}$ per share

11. A total of $16\frac{2}{3}$ gallons of water were drained from a tank holding $37\frac{1}{2}$ gallons. How many gallons were left in the tank?

$20\frac{5}{6}$ gallons

12. Miguel drives $216\frac{1}{2}$ miles on the first leg of a trip and $173\frac{3}{8}$ miles on the second leg. If the total mileage for the trip is $516\frac{3}{4}$ miles, how much farther does he have to travel?

$126\frac{7}{8}$ miles

13. A condominium has $750\frac{7}{8}$ square feet of living space. The master bedroom occupies an area of $172\frac{1}{2}$ square feet and the second bedroom occupies 144 square feet. How much area of living space is left in the condominium?

$434\frac{3}{8}$ square feet

14. Nancy mistakenly washed a cotton tablecloth in hot water. If the tablecloth originally measured $96\frac{1}{2}$ inches and after washing it measured $95\frac{3}{5}$ inches, how much did it shrink?

$\frac{9}{10}$ of an inch

15. Bob and Rita purchased a piece of lakefront property. The sides of the lot measured $112\frac{3}{4}$ feet, $236\frac{9}{10}$ feet, $79\frac{3}{4}$ feet, and $66\frac{2}{5}$ feet. What is the total length of the boundaries of their property?

$495\frac{4}{5}$ feet

16. Stephanie purchased $2\frac{1}{2}$ pounds of ham, 3 pounds of roast beef, $1\frac{1}{4}$ pounds of swiss cheese, $4\frac{3}{4}$ pounds of potato salad, and $2\frac{3}{8}$ pounds of cole slaw for a family barbecue. What was the total weight of her purchase?

$13\frac{7}{8}$ pounds

17. Phil spends $4\frac{2}{3}$ hours at his health club on Thursdays. He lifts weights for $1\frac{1}{2}$ hours and plays racquetball for $1\frac{3}{5}$ hours. What is the total time that remains?

$1\frac{17}{30}$ hours

18. A trip to western Massachussetts from Boston takes $2\frac{2}{3}$ hours. If Cathy drove for $1\frac{1}{2}$ hours, stopped for coffee, and resumed driving for $\frac{3}{4}$ of an hour, how much longer does she have to travel?

$\frac{5}{12}$ hour

19. The price of Bargain Basement stock has risen from $15\frac{7}{8}$ per share on Tuesday to $17\frac{3}{4}$ on Thursday. What is the difference between the stock's sale price on Tuesday and its price on Thursday?

$1\frac{7}{8}$ per share

20. Pat receives 24 vacation days per year. Thus far this year she has used $5\frac{1}{2}$ days in January, 3 days in March, $1\frac{1}{8}$ days in April, and $3\frac{1}{4}$ days in May. How many vacation days does Pat have left?

$11\frac{1}{8}$ days

UNIT

27 Multiplying Fractions

Multiplying fractions is a straightforward procedure because common denominators are unnecessary. When multiplying fractions, numerators must be multiplied by numerators and denominators must be multiplied by denominators. Also, the technique of **cancellation** can be used to reduce fractions within the problem instead of having to reduce the final product.

Cancellation is the process of repeatedly dividing any numerator and any denominator of the problem by the largest possible number that divides both evenly. Again, doing this is just another way of reducing before multiplying to get an answer. Follow these steps when multiplying fractions:

1. Use cancellation if possible.
2. Multiply numerators.
3. Multiply denominators.
4. Reduce to lowest terms if cancellation wasn't used.

EXAMPLE 1 **Multiply:** $\frac{7}{9} \times \frac{6}{14}$.

$$\frac{\overset{1}{\cancel{7}}}{\underset{3}{\cancel{9}}} \times \frac{\overset{\overset{1}{\cancel{2}}}{\cancel{6}}}{\underset{\underset{1}{\cancel{2}}}{\cancel{14}}}$$

Use cancellation:

1. Cancel 7 and 14 by 7.
2. Cancel 6 and 9 by 3.
3. Cancel 2 and 2 by 2.

Answer: $\frac{1 \times 1}{3 \times 1} = \frac{1}{3}$

4. Multiply numerators (1 × 1).
5. Multiply denominators (3 × 1).

PRACTICE 1 **Multiply:** $\frac{3}{16} \times \frac{8}{27}$.

Answer: $\frac{1}{18}$

EXAMPLE 2 **Multiply:** $\frac{6}{7} \times \frac{8}{15} \times \frac{28}{32}$.

$$\frac{\overset{2}{\cancel{6}}}{\underset{1}{\cancel{7}}} \times \frac{\overset{1}{\cancel{8}}}{\underset{5}{\cancel{15}}} \times \frac{\overset{\overset{1}{\cancel{4}}}{\cancel{28}}}{\underset{\underset{1}{\cancel{4}}}{\cancel{32}}}$$

Use cancellation:

1. Cancel 7 and 28 by 7.
2. Cancel 8 and 32 by 8.
3. Cancel 6 and 15 by 3.
4. Cancel 4 and 4 by 4.
5. Multiply numerators (2 × 1 × 1).
6. Multiply denominators (1 × 5 × 1).

Answer: $\frac{2 \times 1 \times 1}{1 \times 5 \times 1} = \frac{2}{5}$

PRACTICE 2 **Multiply:** $\frac{4}{27} \times \frac{12}{14} \times \frac{7}{11}$.

Answer: $\frac{8}{99}$

UNIT **27** Name ____________ Date ____________

Multiplying Fractions

Multiply:

1. $\frac{2}{3} \times \frac{7}{8} =$ $\frac{7}{12}$
2. $\frac{3}{4} \times \frac{2}{5} =$ $\frac{3}{10}$
3. $\frac{7}{8} \times \frac{2}{3} =$ $\frac{7}{12}$
4. $\frac{2}{3} \times \frac{2}{5} =$ $\frac{4}{15}$
5. $\frac{1}{2} \times \frac{7}{8} =$ $\frac{7}{16}$
6. $\frac{3}{4} \times \frac{3}{4} =$ $\frac{9}{16}$
7. $\frac{3}{4} \times \frac{8}{9} =$ $\frac{2}{3}$
8. $\frac{5}{6} \times \frac{12}{25} =$ $\frac{2}{5}$
9. $\frac{8}{15} \times \frac{27}{32} =$ $\frac{9}{20}$
10. $\frac{14}{36} \times \frac{14}{7} =$ $\frac{7}{9}$
11. $\frac{6}{10} \times \frac{5}{3} =$ 1
12. $\frac{7}{12} \times \frac{6}{5} =$ $\frac{7}{10}$
13. $\frac{14}{15} \times \frac{6}{7} =$ $\frac{4}{5}$
14. $\frac{2}{3} \times \frac{1}{4} \times \frac{3}{5} =$ $\frac{1}{10}$
15. $\frac{4}{10} \times \frac{6}{12} \times \frac{1}{8} =$ $\frac{1}{40}$
16. $\frac{2}{3} \times \frac{4}{5} \times \frac{6}{7} =$ $\frac{16}{35}$
17. $\frac{5}{9} \times \frac{9}{4} \times \frac{8}{15} =$ $\frac{2}{3}$
18. $\frac{27}{32} \times \frac{15}{33} \times \frac{4}{35} =$ $\frac{27}{616}$
19. $\frac{6}{42} \times \frac{9}{15} \times \frac{7}{14} =$ $\frac{3}{70}$
20. $\frac{13}{24} \times \frac{16}{26} \times \frac{8}{12} =$ $\frac{2}{9}$
21. $\frac{3}{7} \times \frac{5}{11} \times \frac{22}{25} =$ $\frac{6}{35}$
22. $\frac{32}{81} \times \frac{27}{72} \times \frac{3}{12} =$ $\frac{1}{27}$
23. $\frac{5}{14} \times \frac{3}{9} \times \frac{7}{12} =$ $\frac{5}{72}$
24. $\frac{5}{15} \times \frac{7}{35} \times \frac{3}{4} \times \frac{6}{42} =$ $\frac{1}{140}$
25. $\frac{100}{150} \times \frac{25}{90} \times \frac{4}{5} =$ $\frac{4}{27}$
26. $\frac{26}{34} \times \frac{6}{18} \times \frac{3}{4} \times \frac{5}{12} =$ $\frac{65}{816}$
27. $\frac{8}{15} \times \frac{30}{60} \times \frac{27}{88} \times \frac{15}{24} =$ $\frac{9}{176}$
28. $\frac{9}{21} \times \frac{3}{10} \times \frac{16}{21} \times \frac{5}{12} =$ $\frac{2}{49}$

Business Applications:

29. In a sales office, $\frac{2}{3}$ of the sales professionals are female. Of these female sales professionals, $\frac{3}{4}$ of them had prior experience. Multiply these fractions to find what fraction of the sales force are experienced females.

 $\frac{1}{2}$ *of the sales force*

30. From a $\frac{3}{4}$ ton shipment of bark mulch, $\frac{1}{4}$ is set aside and not available for sale to the public. Multiply to find the fraction of the ton that has been set aside.

 $\frac{3}{16}$ *of a ton*

UNIT

28 Multiplying Whole Numbers, Mixed Numbers, and Fractions

When multiplying whole numbers, mixed numbers, and fractions:

1. Rewrite any whole numbers as fractions with denominators of 1.
2. Change any mixed numbers to improper fractions.
3. Use cancellation if possible.
4. Multiply the numerators.
5. Multiply the denominators.
6. Reduce to lowest terms if cancellation wasn't used. Change any improper fractions to mixed numbers.

EXAMPLE 1 **Multiply: $8 \times 5\frac{1}{3}$.**

Answer: $8 \times 5\frac{1}{3} = \frac{8}{1} \times \frac{16}{3} = \frac{128}{3} = 42\frac{2}{3}$

PRACTICE 1 **Multiply: $6 \times 7\frac{1}{2}$.**

Answer: 45

EXAMPLE 2 **Multiply: $4\frac{1}{3} \times 2\frac{2}{5} \times \frac{1}{6}$.**

Answer: $4\frac{1}{3} \times 2\frac{2}{5} \times \frac{1}{6} = \frac{13}{3} \times \frac{\overset{2}{\cancel{12}}}{5} \times \frac{1}{\underset{1}{\cancel{6}}} = \frac{13 \times 2 \times 1}{3 \times 5 \times 1} = \frac{26}{15} = 1\frac{11}{15}$

PRACTICE 2 **Multiply: $3\frac{3}{4} \times 1\frac{2}{5} \times \frac{12}{19}$.**

Answer: $3\frac{6}{19}$

UNIT 28 Name ______________________ Date ____________

Multiplying Whole Numbers, Mixed Numbers, and Fractions

Multiply the following:

1. $\frac{1}{8} \times 7\frac{1}{2} =$ $\frac{15}{16}$
2. $\frac{5}{6} \times 4\frac{1}{3} =$ $3\frac{11}{18}$
3. $\frac{3}{5} \times 3\frac{3}{4} =$ $2\frac{1}{4}$
4. $\frac{4}{5} \times 3\frac{3}{4} =$ 3
5. $\frac{5}{8} \times 1\frac{3}{5} =$ 1

6. $\frac{7}{9} \times 3\frac{3}{7} =$

$2\frac{2}{3}$

7. $6 \times 4\frac{3}{8} =$

$26\frac{1}{4}$

8. $8 \times 4\frac{6}{7} =$

$38\frac{6}{7}$

9. $9 \times 6\frac{1}{3} =$

57

10. $7 \times 5\frac{7}{8} =$

$41\frac{1}{8}$

11. $12 \times 5\frac{5}{16} =$

$63\frac{3}{4}$

12. $6 \times 2\frac{2}{15} =$

$12\frac{4}{5}$

13. $2\frac{2}{9} \times 4\frac{1}{3} =$

$9\frac{17}{27}$

14. $7\frac{3}{7} \times 4\frac{1}{6} =$

$30\frac{20}{21}$

15. $8\frac{2}{3} \times 16\frac{2}{4} =$

143

16. $1\frac{9}{11} \times 1\frac{1}{16} =$

$1\frac{41}{44}$

17. $2\frac{3}{7} \times 3\frac{1}{9} =$

$7\frac{5}{9}$

18. $4\frac{3}{7} \times 2\frac{5}{8} =$

$11\frac{5}{8}$

19. $4\frac{3}{10} \times 2\frac{7}{10} \times 12 =$

$139\frac{8}{25}$

20. $3\frac{1}{2} \times 4\frac{1}{7} \times 16 =$

232

21. $12 \times 1\frac{1}{2} \times 4\frac{3}{9} =$

78

22. $16\frac{2}{3} \times \frac{7}{8} \times 5 =$

$72\frac{11}{12}$

23. $3\frac{8}{13} \times 5\frac{1}{5} \times 14 =$

$263\frac{1}{5}$

24. $11\frac{2}{3} \times 14 \times 1\frac{2}{5} =$

$228\frac{2}{3}$

25. $13\frac{1}{6} \times \frac{2}{3} \times 4 \times \frac{2}{7} =$

$10\frac{2}{63}$

26. $16 \times 4\frac{2}{3} \times \frac{3}{8} \times 1\frac{1}{10} =$

$30\frac{4}{5}$

27. $\frac{5}{16} \times \frac{3}{12} \times 4\frac{2}{5} \times 6 =$

$2\frac{1}{16}$

Business Applications:

28. In an insurance office, $\frac{2}{3}$ of the 351 claims filed last month were auto related. Multiply these quantities to arrive at the number of auto claims filed.

234 auto claims

29. A $6\frac{1}{2}$-acre lot of land is $\frac{2}{3}$ wooded. Multiply these fractions to find the number of wooded acres.

$4\frac{1}{3}$ acres

UNIT

29 Dividing Fractions

Dividing fractions requires the introduction of a new term—**reciprocal.** The reciprocal of a number is the number that when multiplied times the original gives the product 1.

For example, $\frac{1}{3}$ is the reciprocal of 3 because

$$\frac{1}{3} \times 3 = \frac{3}{3} = 1\,.$$

The reciprocal of a fraction is found by interchanging the numerator with the denominator. This process is often referred to as **inverting** the fraction.

For example, the reciprocal of $\frac{7}{9}$ is $\frac{9}{7}$.

This process of inverting or writing the reciprocal is used when dividing numbers because division means "multiplied by the reciprocal."

For example, the problem:

$$16 \div 4 = 4$$

can be rewritten as

$$16 \times \frac{1}{4} = \frac{16}{4} = 4\,.$$

By doing this, we see that division can be done by replacing the division symbol with multiplication and writing the reciprocal of the divisor.

Follow these steps when dividing fractions:

1. Replace the division symbol with a multiplication symbol and invert the divisor (write the reciprocal of the divisor). The fraction to be inverted is the one to the immediate right of the division symbol.
2. Use cancellation whenever possible.
3. Multiply numerators.
4. Multiply denominators.
5. Reduce to lowest terms or change to a mixed number if necessary.

EXAMPLE 1 **Find the reciprocal of the following:**

Answers: a. 6 The reciprocal of 6 is $\frac{1}{6}$. b. $\frac{3}{7}$ The reciprocal of $\frac{3}{7}$ is $\frac{7}{3}$.

PRACTICE 1 **Find the reciprocal of 13:**

Answer: $\frac{1}{13}$

EXAMPLE 2 **Divide: $\frac{6}{18} \div \frac{2}{15}$.**

Answer: $\frac{6}{18} \div \frac{2}{15} = \frac{\overset{1}{\cancel{\overset{3}{\cancel{6}}}}}{\underset{2}{\cancel{\underset{6}{\cancel{18}}}}} \times \frac{\overset{5}{\cancel{15}}}{\underset{1}{\cancel{2}}} = \frac{5}{2} = 2\frac{1}{2}$

PRACTICE 2 **Divide:** $\frac{4}{9} \div \frac{12}{21}$.

Answer: $\frac{7}{9}$

UNIT **29** Name ______________________ Date ______________

Dividing Fractions

Find the reciprocal of the following:

1. $\frac{4}{9} = \frac{9}{4}$
2. $\frac{2}{3} = \frac{3}{2}$
3. $5 = \frac{1}{5}$
4. $\frac{1}{13} = 13$

Divide:

5. $\frac{2}{3} \div \frac{3}{4} =$ $\frac{8}{9}$
6. $\frac{4}{7} \div \frac{7}{7} =$ $\frac{4}{7}$
7. $\frac{1}{4} \div \frac{3}{8} =$ $\frac{2}{3}$
8. $\frac{5}{9} \div \frac{8}{36} =$ $2\frac{1}{2}$
9. $\frac{5}{6} \div \frac{1}{3} =$ $2\frac{1}{2}$
10. $\frac{5}{9} \div \frac{2}{5} =$ $1\frac{7}{18}$
11. $\frac{9}{16} \div \frac{3}{4} =$ $\frac{3}{4}$
12. $\frac{3}{7} \div \frac{15}{28} =$ $\frac{4}{5}$
13. $\frac{3}{5} \div \frac{9}{10} =$ $\frac{2}{3}$
14. $\frac{1}{3} \div \frac{3}{5} =$ $\frac{5}{9}$
15. $\frac{4}{5} \div \frac{3}{10} =$ $2\frac{2}{3}$
16. $\frac{5}{6} \div \frac{15}{18} =$ 1
17. $\frac{3}{4} \div \frac{3}{8} =$ 2
18. $\frac{5}{9} \div \frac{5}{3} =$ $\frac{1}{3}$
19. $\frac{5}{28} \div \frac{25}{42} =$ $\frac{3}{10}$
20. $\frac{7}{11} \div \frac{14}{33} =$ $1\frac{1}{2}$
21. $\frac{11}{60} \div \frac{44}{50} =$ $\frac{5}{24}$
22. $\frac{18}{7} \div \frac{9}{21} =$ 6
23. $\frac{33}{7} \div \frac{11}{21} =$ 9
24. $\frac{6}{11} \div \frac{2}{3} =$ $\frac{9}{11}$

Business Applications:

25. A bag of galvanized deck screws weighs $\frac{3}{4}$ pound. If one screw weighs $\frac{2}{64}$ of a pound, divide these quantities to find how many screws are in the $\frac{3}{4}$-pound bag.

    24 screws

26. If over a $\frac{7}{8}$-mile stretch of highway, markers are to be placed $\frac{1}{16}$ of a mile apart, divide to find how many $\frac{1}{16}$-mile sections will be created over this distance.

    14 sections

UNIT

30 Dividing Whole Numbers, Mixed Numbers, and Fractions

When dividing whole numbers, mixed numbers, and fractions:

1. Rewrite any whole numbers as fractions with denominators of 1.
2. Change any mixed numbers to improper fractions.
3. Change division to multiplication and invert the divisor.
4. Use cancellation if possible.
5. Multiply the numerators.
6. Multiply the denominators.
7. Reduce to lowest terms or change to a mixed number if necessary.

EXAMPLE 1 **Divide: $\frac{3}{4} \div 6$.**

Answer: $\frac{3}{4} \div 6 = \frac{3}{4} \div \frac{6}{1} = \frac{\overset{1}{\cancel{3}}}{4} \times \frac{1}{\underset{2}{\cancel{6}}} = \frac{1}{8}$

PRACTICE 1 **Divide: $\frac{5}{9} \div 5$.**

Answer: $\frac{1}{9}$

EXAMPLE 2 **Divide: $4\frac{1}{3} \div 2\frac{1}{4}$.**

Answer: $4\frac{1}{3} \div 2\frac{1}{4} = \frac{13}{3} \div \frac{9}{4} = \frac{13}{3} \times \frac{4}{9} = \frac{52}{27} = 1\frac{25}{27}$

PRACTICE 2 **Divide: $2\frac{3}{8} \div 1\frac{1}{4}$.**

Answer: $1\frac{9}{10}$

UNIT 30 Name ______________________ Date ______________

Dividing Whole Numbers, Mixed Numbers, and Fractions

Divide:

1. $\frac{7}{8} \div 2\frac{3}{4} =$

   $\frac{7}{22}$

2. $8\frac{1}{2} \div 2\frac{1}{4} =$

   $3\frac{7}{9}$

3. $6\frac{1}{4} \div 4\frac{1}{3} =$

   $1\frac{23}{52}$

4. $6\frac{1}{10} \div 4\frac{3}{2} =$

   $1\frac{6}{55}$

5. $\frac{2}{5} \div 4\frac{2}{7} =$

   $\frac{7}{75}$

6. $7\frac{3}{7} \div 2\frac{1}{14} =$

   $3\frac{17}{29}$

7. $3\frac{1}{4} \div \frac{1}{3} =$

   $9\frac{3}{4}$

8. $4 \div 2\frac{1}{9} =$

   $1\frac{17}{19}$

9. $2\frac{1}{4} \div 3 =$

   $\frac{3}{4}$

10. $1\frac{1}{10} \div 5 =$

    $\frac{11}{50}$

11. $12\frac{1}{6} \div 3 =$

    $4\frac{1}{18}$

12. $12 \div \frac{3}{14} =$

    56

13. $6 \div \frac{4}{7} =$

    $10\frac{1}{2}$

14. $5\frac{3}{4} \div \frac{3}{4} =$

    $7\frac{2}{3}$

15. $4 \div 2\frac{1}{2} =$

    $1\frac{3}{5}$

16. $5\frac{1}{4} \div 6 =$

    $\frac{7}{8}$

17. $\frac{9}{10} \div 1\frac{1}{15} =$

    $\frac{27}{32}$

18. $4\frac{3}{4} \div 2\frac{3}{4} =$

    $1\frac{8}{11}$

19. $1\frac{3}{4} \div 9 =$

    $\frac{7}{36}$

20. $\frac{3}{7} \div 1\frac{19}{21} =$

    $\frac{9}{40}$

21. $6 \div 2\frac{2}{3} =$

    $2\frac{1}{4}$

22. $1\frac{2}{3} \div 1\frac{1}{9} =$

    $1\frac{1}{2}$

23. $12 \div 3\frac{1}{3} =$

    $3\frac{3}{5}$

24. $\frac{1}{4} \div \frac{2}{4} =$

    $\frac{1}{2}$

Business Applications:

25. A 900-acre lot is to be subdivided into $2\frac{1}{4}$-acre housing lots. Divide these numbers to find how many housing lots can be obtained from this parcel of land.

    400 lots

26. A produce market has received 620 crates of apples and plans to bag these apples in plastic. Each bag can hold $\frac{1}{8}$ of a crate. Divide to find how many bags of apples can be prepared.

    4,960 bags

How to Dissect and Solve Word Problems: Multiplication and Division of Fractions

EXAMPLE 1 **In a class of 33 students, $\frac{2}{3}$ are boys. How many boys are in the class?**

Gather the facts.	What am I solving for?	What must I need to know or calculate before solving the problem?	Key points to remember.
Total students = 33 Boys = $\frac{2}{3}$ of 33	Number of boys in the class	Total students × $\frac{2}{3}$ = Number of boys in the class	Use cancellation when possible.

Answer: $33 \times \frac{2}{3} = \frac{33}{1} \times \frac{2}{3} = 22$ boys

PRACTICE 1 **Out of a class of 36 students, $\frac{3}{4}$ are girls. How many girls are in the class?**

Answer: 27 girls

EXAMPLE 2 **Sheila participated in a road race of $6\frac{1}{4}$ miles. She jogged $\frac{4}{5}$ of the total distance and for $\frac{1}{2}$ of this distance she was alone. How many miles did she jog alone?**

Gather the facts.	What am I solving for?	What must I need to know or calculate before solving the problem?	Key points to remember.
Total miles = $6\frac{1}{4}$ Jogging miles = $\frac{4}{5}$ of total distance Alone = $\frac{1}{2}$ of jogging distance	How many miles did she jog alone?	Total miles × $\frac{4}{5}$ × $\frac{1}{2}$ = Miles jogged alone	Change mixed numbers to improper fractions. Cancel when possible. Reduce answer if necessary.

Answer: $6\frac{1}{4} \times \frac{4}{5} \times \frac{1}{2} = \frac{25}{4} \times \frac{4}{5} = \frac{5}{1} \times \frac{1}{2} = \frac{5}{2} = 2\frac{1}{2}$ miles

PRACTICE 2 **Two-fifths of a shipment of 625 sneakers were defective. Of these defective items, $\frac{3}{5}$ of them can be sold as seconds at a discount store. How many sneakers can be sold as seconds?**

Answer: 150 sneakers

EXAMPLE 3 **The net weight of a box of hardware is $4\frac{2}{5}$ pounds. Each item of hardware weighs $\frac{11}{65}$ pounds. How many items are in the box?**

Gather the facts.	What am I solving for?	What must I need to know or calculate before solving the problem?	Key points to remember.
Net weight = $4\frac{2}{5}$ pounds Each item = $\frac{11}{65}$ pounds	Number of items in box	Net weight ÷ Weight of each item = Number of items in box	Change mixed numbers to improper fractions. Invert divisor and multiply. Reduce answer if necessary.

Answer: $4\frac{2}{5} \div \frac{11}{65} = \frac{22}{5} \div \frac{11}{65} = \frac{22}{5} \times \frac{65}{11} = 26$ items

PRACTICE 3 **Each individually wrapped package of medication weighs $\frac{3}{7}$ pounds. How many packages of medication are there if the carton they are shipped in weighs 21 pounds?**

Answer: 49 packages

UNIT **31** Name ______________________ Date ______________

How to Dissect and Solve Word Problems: Multiplication and Division of Fractions

Solve the following problems:

1. A wall covering is $1\frac{3}{4}$ inches thick. If $\frac{2}{7}$ of this thickness is plasterboard, how thick is the plasterboard?

   $\frac{1}{2}$ inch

2. Kevin drives at a constant rate of $1\frac{1}{3}$ miles per minute. How many miles does he travel in $3\frac{3}{4}$ minutes?

   5 miles

3. A tin container holds $5\frac{1}{4}$ pounds of buttons. How many pounds of buttons are there in $2\frac{1}{2}$ tin containers?

   $13\frac{1}{8}$ pounds

4. How many $2\frac{1}{2}$-inch pieces of wood can be cut from a length of wood 30 inches long?

   12 pieces

5. If Joe can haul $\frac{3}{4}$ tons of sand in his truck, how many truckloads will it require to move $7\frac{1}{2}$ tons of sand?

   10 truckloads

6. Find the proposed increase in tuition if the increase is to be $\frac{1}{20}$ of the current tuition of $3,500.

   $175

7. After a party, $\frac{1}{5}$ of $7\frac{1}{2}$ pounds of meat purchased for the barbecue was left. Of this amount, $\frac{1}{4}$ was hamburger meat. How many pounds of hamburger meat were left over?

   $\frac{3}{8}$ pound

8. If a plane flies $45\frac{1}{2}$ miles in $6\frac{1}{2}$ minutes, how many miles does it fly per minute?

   7 miles

9. In a class of 42 students, $\frac{5}{7}$ are males. How many females are in the class?

   12 females

10. Peter can shovel his driveway in $3\frac{3}{4}$ hours. Using a snowblower he can complete the job in $\frac{2}{3}$ of the time. How long will it take to clear the driveway with the snowblower?

    $2\frac{1}{2}$ hours

11. If a car is driven $120\frac{3}{4}$ miles in $2\frac{1}{2}$ hours, what is the average hourly rate?

$48\frac{3}{10}$ miles per hour

12. Amber donates $\frac{1}{10}$ of her salary to charitable causes. If her salary is $72,000 per year and $\frac{2}{3}$ of her charitable contribution goes to organizations for the homeless, how much does she give to these organizations?

$4,800

13. If each ski run Mike takes lasts $3\frac{3}{4}$ minutes, how many runs can he make in 60 minutes?

16 runs

14. Diane is wrapping Christmas presents, and she has a ribbon measuring $\frac{7}{9}$ of a yard long. She wants to cut this into 7 pieces of equal length. How many inches will each piece be? (Note: 1 yard = 36 inches.)

4 inches

15. On a map, 1 mile is represented by $\frac{3}{4}$ inches. How many miles are represented by a line $8\frac{5}{8}$ inches long?

$11\frac{1}{2}$ miles

16. A barrel contains $\frac{17}{18}$ of a ton of sand. If $\frac{1}{36}$ of a ton is used each day, how many days will the sand last?

34 days

17. How many slices of pound cake can be cut from a cake $13\frac{3}{4}$ inches long if each piece is $\frac{5}{8}$ inches thick?

22 slices

18. In a $7\frac{1}{2}$-hour workday, Christina spends $\frac{2}{3}$ of her time engaged in office procedures. Three-eighths of this time is spent on computer related tasks. How many hours are spent on computer tasks?

$1\frac{7}{8}$ hours

19. How many $2\frac{1}{2}$-inch tiles are needed to fill a space $8\frac{3}{4}$ feet long?

42 tiles

20. Arthur advanced to the final round in $\frac{7}{13}$ of the 26 tennis tournaments he played in this year. He won $\frac{4}{7}$ of these final matches. How many tournaments did Arthur win?

8 tournaments

CHAPTER 2

Name ____________________ Date ____________

Test

Answers

1. Identify the following as proper fractions (*P*), improper fractions (*I*), mixed numbers (*M*), or fractions that equal 1 (1).

   a. $\frac{7}{6}$ b. $\frac{13}{13}$ c. $\frac{27}{4}$ d. $\frac{1}{9}$ e. $13\frac{3}{8}$

1a. I

1b. 1

1c. I

1d. P

1e. M

2. Change to a mixed number: $\frac{46}{7} =$

2. $6\frac{4}{7}$

3. Change to an improper fraction: $6\frac{3}{5} =$

3. $\frac{33}{5}$

4. Reduce to lowest terms:

   a. $\frac{14}{42} =$ b. $3\frac{16}{64} =$

4a. $\frac{1}{3}$

4b. $3\frac{1}{4}$

5. Raise to higher terms: $\frac{3}{13} = \frac{?}{52}$

5. $\frac{12}{52}$

6. Add: $\frac{3}{5} + \frac{1}{2} + \frac{9}{10} =$

6. 2

7. Add: $6\frac{5}{16} + \frac{3}{4} + 2\frac{1}{3} =$

7. $9\frac{19}{48}$

(Test continues on next page)

CHAPTER 2

Test *(Concluded)*

Answers

8. Subtract: $\frac{7}{8} - \frac{2}{6} =$

8. $\frac{13}{24}$

9. Subtract: $10\frac{1}{4} - 3\frac{4}{5} =$

9. $6\frac{9}{20}$

10. An 8-ton load of crushed stone is being used at a housing construction site. Two and a half tons are being used as part of a cement mixture for a floor. Another $1\frac{3}{8}$ tons are being used as part of a cement mixture for building a chimney. The remaining stone will be used around the perimeter of the house's foundation. How many tons of stone remain for this purpose?

10. $4\frac{1}{8}$ tons

11. Multiply: $\frac{22}{27} \times \frac{9}{16} \times 4\frac{2}{11} =$

11. $1\frac{11}{12}$

12. Divide: $3\frac{7}{15} \div 2\frac{3}{5} =$

12. $1\frac{1}{3}$

13. A 207-acre site is to be developed for housing. Two-thirds of the entire site will be used for the houses. This portion will then be subdivided into $1\frac{1}{2}$-acre housing lots. How many housing lots will be obtained by this subdivision?

13. 92 lots

CHAPTER

3 Decimals

UNITS

UNIT

32 Writing Decimal Numbers in Verbal and Numeral Form

Decimals exist as another way of expressing the value of any common fraction. Digits along with a period, called the *decimal point*, are used to express these quantities. Numbers expressed this way are written in *decimal form*. As with whole numbers, numbers written in decimal form use a system of place values. The diagram below illustrates this place value system.

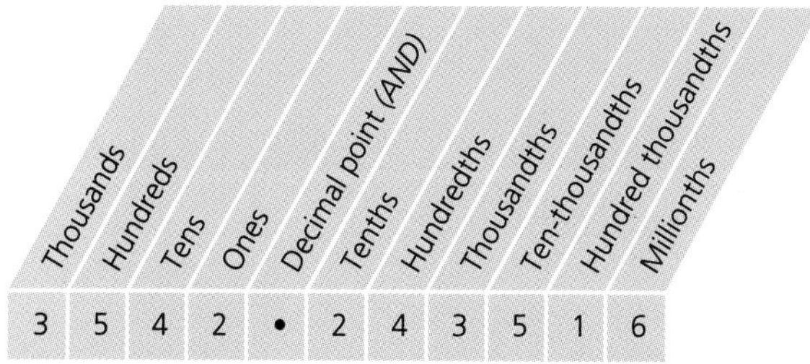

The digits to the left of the decimal point express the whole number using the same place value system as before. Digits to the right of the decimal point express the fractional portion of the number and are referred to as the *decimal parts*. The number in the above illustration is three thousand five hundred forty-two and two hundred forty-three thousand, five hundred sixteen millionths.

It is important to understand that the decimal point itself does not express a place value. It is only a point of separation between the whole number and the decimal parts. It is read aloud as *and*. Numbers written using this notation are simply referred to as **decimals.**

The relationship of the numbers on either side of the decimal point should be made clear at this point. If the decimal point is moved one place to the right, the number's value is *increased* ten times. If the decimal point is moved one place to the left, the number's value is *decreased* ten times. Consequently, the decimal point is often thought of as being the central part of the decimal number system.

To read a decimal number aloud:

1. Find the decimal point.
2. If necessary, read the number to the left of the decimal point as a whole number.
3. If necessary, say *and* for the decimal point.
4. Read the number to the right of the decimal point and add the name of the place of the last digit.

To write a decimal number in verbal form:

1. If necessary, write the name for the whole number to the left of the decimal point.
2. If necessary, write the word *and* for the decimal point.
3. Write the whole number name for the number to the right of the decimal point.
4. Write the name of the place value of the last digit.

To write a decimal number in numeral form:

1. If necessary, write the whole number followed by a decimal point to replace the word *and*.
2. Find how many places are needed to write the decimal part of the number.
3. If needed, use as many zeros as necessary to fill spaces to the left of the digits written to express the decimal part of the number.

EXAMPLE 1 **Write the following decimals in words.**

Answers: .049 forty-nine thousandths

13.076 thirteen and seventy-six thousandths

PRACTICE 1 **Write the following decimals in words.**

Answers: .0016 sixteen ten thousandths

41.017 forty-one and seventeen thousandths

EXAMPLE 2 **Write the following as decimal numbers.**

Answers: thirty-seven thousandths .037

ninety-five and sixty-eight ten thousandths 95.0068

PRACTICE 2 **Write the following as decimal numbers.**

Answers: eighty-one ten thousandths .0081

fifty and ninety-three hundredths 50.93

UNIT **32** Name ______________________ Date ______________

Writing Decimal Numbers in Verbal and Numeral Form

Write the following as decimal numbers:

1. three tenths .3
2. forty-three and six hundredths 43.06
3. twenty-nine hundredths .29
4. seventeen and eighty-nine thousandths 17.089
5. fifteen thousandths .015
6. five and nine hundred seventy-two ten thousandths 5.0972
7. forty-three ten thousandths .0043
8. nine hundred fifty-seven ten thousandths .0957
9. forty-nine and sixty-five ten thousandths 49.0065

UNIT 32

Name ______________________ Date ______________

Reading and Writing Decimal Numbers in Verbal and Numeral Form

10. ten and forty-four hundredths — 10.44
11. three hundred thousandths — .300
12. twenty and two ten thousandths — 20.0002
13. one hundred and five thousandths — 100.005

Write the following decimals in words:

14. .9 — nine tenths
15. .53 — fifty-three hundredths
16. .407 — four hundred seven thousandths
17. 4.07 — four and seven hundredths
18. 62.014 — sixty-two and fourteen thousandths
19. 300.3 — three hundred and three tenths
20. .0069 — sixty-nine ten thousandths
21. 15.015 — fifteen and fifteen thousandths
22. .0637 — six hundred thirty-seven ten thousandths
23. .00006 — six hundred-thousandths
24. 10.0007 — ten and seven ten thousandths
25. 87.5431 — eighty-seven and five thousand four hundred thirty-one ten thousandths
26. 333.33 — three hundred thirty-three and thirty-three hundredths

Business Applications:

27. The parking lot of a national food wholesaler covers two and seven thousandths acres. Express this quantity as a decimal number.

    2.007

28. Labor Department statistics reveal the average factory workweek to be 41.2 hours. Express this decimal in words.

    forty-one and two tenths

Rounding Decimals

Rounding decimal numbers is very similar to rounding whole numbers.

1 Identify the digit you wish to round.

2 If the digit to the right of the digit to be rounded is 5 or more, increase the identified digit by 1. If the digit is less than 5, do not change the identified digit.

3 Drop all digits to the right of the identified digit.

EXAMPLE 1 **Round to the nearest tenth.**

Answers: .13 __.1__ .27 __.3__ .0532 __.1__ 16.2745 __16.3__

PRACTICE 1 **Round to the nearest hundredth.**

Answers: .174 __.17__ .839 __.84__ 5.6075 __5.61__ 14.2919 __14.29__

EXAMPLE 2 **Round the following to the given place.**

Round to the nearest tenth: 44.98 **Answer:** __45.0__

Round to the nearest cent: $61.475 **Answer:** __$61.48__

Round to the nearest dollar: $23.49 **Answer:** __$23.00__

PRACTICE 2 **Round the following to the given place.**

Round to the nearest tenth: 17.95 **Answer:** __18.0__

Round to the nearest cent: $33.768 **Answer:** __$33.77__

Round to the nearest dollar: $119.56 **Answer:** __$120.00__

UNIT 33 Name ______________________ Date __________

Rounding Decimals

Round the following to the given place:

1.	4.437	tenth	4.4
2.	7.1546	thousandth	7.155
3.	8.75	tenth	8.8
4.	42.889	tenth	42.9
5.	500.025	hundredth	500.03
6.	81.119	hundredth	81.12
7.	.102367	ten thousandth	.1024
8.	.345	tenth	.3
9.	4.8715	hundredth	4.87
10.	.99	tenth	1.0
11.	4.186	hundredth	4.19
12.	.36229	thousandth	.362
13.	.7358	thousandth	.736
14.	4.895	hundredth	4.90
15.	.6721	tenth	.7
16.	.2408	hundredth	.24
17.	1.7430	thousandth	1.743
18.	14.553	tenth	14.6
19.	3.0007	thousandth	3.001
20.	10.0752	hundredth	10.08
21.	36.58	nearest whole number	37
22.	$100.495	nearest cent	$100.50
23.	$78.91	nearest dollar	$79.00
24.	$5.39	nearest dollar	$5.00
25.	15.5	nearest whole number	16
26.	3,700.89	nearest whole number	3,701
27.	$3,049.50	nearest dollar	$3,050
28.	$965.078	nearest cent	$965.08

Business Applications:

29. An oil tank manufacturer states the capacity of its standard tank to be 506.75 gallons. Round this figure to the nearest tenth.

506.8 gallons

30. A carpet retailer measures a room and finds its size to be 104.8 square yards. Round this decimal to the nearest whole number.

105 square yards

UNIT

34 Addition and Subtraction of Decimals

The procedures for adding and subtracting decimal numbers are very similar to the ones used for adding and subtracting whole numbers. Because only numbers having the same place value can be added or subtracted, the decimal points must be lined up with each other when adding and subtracting.

To add decimals:

1. Arrange the numbers to be added vertically so that the decimal points and numbers of the same place value are lined up.
2. Add the numbers in the columns and place the decimal point in the sum directly below the decimal points in the column.

To subtract decimals:

1. Arrange the numbers to be subtracted vertically so that the decimal points and numbers of the same place value are lined up. If necessary, place any zeros in the top number (the minuend) before subtracting.
2. Subtract the numbers, borrowing if necessary.
3. Place the decimal point in the difference directly below the decimal points in the column.

EXAMPLE 1 **Add: 12.37 + 100.489 + .6 + 3.9673.**

```
     12.37
    100.489
       .6
+     3.9673
-----------
    117.4263
```

Answer: 117.4263

PRACTICE 1 **Add: 17.42 + .125 + 207.1279 + 29.009.**

Answer: 253.6819

EXAMPLE 2 **Subtract: 732.46 − 56.4386.**

```
    732.4600
−    56.4386
-----------
    676.0214
```

Answer: 676.0214

PRACTICE 2 **Subtract: 416.37 − 71.5874.**

Answer: 344.7826

UNIT **34** Name ______________________ Date ______________

Addition and Subtraction of Decimals

Add:

1. 19.023 + 9.37 + 5.6 =

   33.993

2. 7.27 + 324.0687 + 14.26 =

   345.5987

3. 2.143 + 5.164 + 2.9 + 139.074 =

   149.281

4. 105.7 + 19.4 + 1,119.05 + 468.006 =

   1,712.156

5. 48.1 + .0481 + 481 + .00481 =

   529.15291

6. 1.151 + 26.23 + 2,002.167 + 13 =

   2,042.548

7. .00369 + 1.07 + 14.123 + .0025 =

   15.19919

8. 390.65 + 49.25 + 2,340.73 + 492.57 =

   3,273.2

9. 91.003 + 16.491 + 160.00471 =

   267.49871

10. 700.05 + 900.0006 + .0315 =

    1,600.0821

11. 61.843 + 120.75 + 142.4056 + .128 =

    325.1266

12. 32.35 + 40.15 + .000763 + 52 =

    124.500763

13. .8 + .87 + .626 + .49628 =

    2.79228

14. 124.2750 + 13.07 + .0012345 =

    137.3462345

15. .017145 + .99 + .017 + 4.121 + 10.4 =

    15.545145

UNIT **34** Name ______________________ Date ______________

Addition and Subtraction of Decimals

Subtract:

16. 21.473 − 16.27 =

5.203

17. 0.685 − .23 =

.455

18. 16.443 − 4.007 =

12.436

19. 3.2319 − 1.567 =

1.6649

20. 35.07 − 16.46 =

18.61

21. 12 − 4.34 =

7.66

22. 617.32 − 4.769 =

612.551

23. 463.05 − 17.0613 =

445.9887

24. 480.17 − 29.457 =

450.713

25. 23.261 − 15.387 =

7.874

26. .145 − .09684 =

.04816

27. 173.37 − 14.75 =

158.62

28. .05 − .000765 =

.049235

Business Applications:

29. A trucking company owns a truck that has a capacity to handle 1,700 pounds. Loaded onto the truck is 450.813 pounds of wood and 699.222 pounds of cement. Combine by adding and subtracting to establish how many pounds the truck is under its capacity.

549.965 pounds

30. Parson's Coffee Shop checking account balance is $3,576.84. At the end of the day $419.36 is deposited into the account. Add these quantities to figure the final balance.

$3,996.20

How to Dissect and Solve Word Problems: Addition and Subtraction of Decimals

Use the chart below as an aid to solving the word problems in this unit.

Gather the facts.	What am I solving for?	What must I need to know or calculate before solving the problem?	Key points to remember.

EXAMPLE 1 **A record of mileage kept over five days revealed that 107.63, 97.3, 216.378, 156.2, and 11.17 miles were driven over that time. What was the total mileage driven (rounded to the nearest hundredth)?**

Gather the facts.	What am I solving for?	What must I need to know or calculate before solving the problem?	Key points to remember.
Mileage recorded: 107.63 97.3 216.378 156.2 11.17	Total mileage	Sum of each day's mileage = Total mileage	Line up decimal points vertically. Hundredths is two places to the right of the decimal point.

Add:

$$\begin{array}{r} 107.63 \\ 97.3 \\ 216.378 \\ 156.2 \\ +\ \ 11.17 \\ \hline 588.678 \end{array}$$

Answer: 588.678 = 588.68 miles

PRACTICE 1 **Alice received four checks for $117.37, $38.96, $142.17, and $23.97 and deposited all of them into her checking account. Her original balance in the account was $839.13. What was the balance after the deposit?**

Answer: $1,161.60

EXAMPLE 2 **Lotus Bookkeeping records its utility expenses of $317.75, $75.07, and $193.10 for one month. If its rent expense is $2,017 per month, what is the difference between its rental and utilities expenses?**

Gather the facts.	What am I solving for?	What must I need to know or calculate before solving the problem?	Key points to remember.
Utilities expenses = $317.25 $75.07 $193.10 Rental expense = $2,017	Difference between rent and utilities	Rental expense − Total utilities expense = Difference between rent and utilities	Line up decimal points. Insert zeros if necessary when subtracting.

Add:
$$\begin{array}{r} \$317.75 \\ \$\ \ 75.07 \\ +\ \$193.10 \\ \hline \$585.92 \end{array}$$

Subtract:
$$\begin{array}{r} \$2{,}017.00 \\ -\ \$\ \ \ 585.92 \\ \hline \$1{,}431.08 \end{array}$$

Answer:

PRACTICE 2 **John spent $1,193, $640.93, and $1,349.72 for house and car insurance last year. Because of a safe driving record and the lowering of house insurance rates, he was reimbursed a total of $310. What did he end up spending on insurance last year?**

Answer: $2873.65

UNIT 35 Name ______________________ Date ______________

How to Dissect and Solve Word Problems: Addition and Subtraction of Decimals

Solve the following problems:

1. The floor space for 4 rooms in a dentist's office measures 233.17 square feet, 119.33 square feet, 272.6 square feet, and 216.58 square feet. What is the total square footage of space?

   841.68 square feet

2. Jean worked 37.3 hours, 40.16 hours, 35.3 hours, and 45.75 hours each of the last 4 weeks. What were the total hours she worked?

   158.51 hours

3. Jocelyn spent $37.16, $73.49, $19.99, and $64.16 on Christmas gifts. What did she spend altogether?

   $194.80

4. A corner lot measures 117 feet, 96.93 feet, 136.3 feet, and 77.25 feet. What is the total length of the sides of the lot?

   427.48 feet

5. The circular area of 3 flower beds measures 272.8862 square feet, 572.5552 square feet, and 321.7743 square feet. What is the total area of the flower beds (rounded to the nearest hundredth)?

   1,167.22 square feet

6. In a relay race 4 runners ran the race in 51.0375, 50.63, 52.1009, and 49.364 seconds respectively. What was the total running time?

   203.1324 seconds

7. A pump discharges 47.3, 46.967, 49, and 50.355 gallons of water each hour for 4 hours. If the total capacity of the tank holding the water is 210.75 gallons, how much water remains after 4 hours?

   17.128 gallons

8. A patient's temperature is recorded at 101.96 degrees at noon. Two hours later, it is recorded as 99.9 degrees. By how many degrees did the patient's temperature drop?

   2.06 degrees

9. Two model racing car replicas measure 24.073 centimeters and 28.42 centimeters respectively. How much longer is the second model than the first?

   4.347 centimeters

10. Theresa cuts 2 lengths of fabric measuring 13.75 yards and 8.875 yards long from a bolt of fabric measuring 30 yards. How many yards of fabric are left on the bolt?

    7.375 yards

11. Four farms containing 61.843 acres, 120.75 acres, 142.4056 acres, and 180.750 acres are merged. What is the total acreage?

    505.7486 acres

12. Thea buys $17.89 and $4.38 worth of food and beverages for a small house party. What is her change from $25.00?

    $2.73

13. A 12.5-foot length of wood trim has to be cut into 3 pieces measuring 3.875 feet, 3.8125 feet, and 2.625 feet. How many feet of trim are left?

    2.1875 feet

14. At noon, the temperature at the top of Mount Washington is 16.8 degrees (Fahrenheit). One hour later the temperature has increased by 1.45 degrees (Fahrenheit). What is the temperature at this time?

    18.25 degrees (Fahrenheit)

15. A checking account at Bank A contains $781.37. The difference between this amount and the amount in a checking account at Bank B is $403.97. How much money is in the checking account at Bank B?

    $377.40

16. Antonio's fluid intake is recorded as 0.475, 0.550, 0.675, and 1.250 liters during his stay at St. Samuel's Hospital. What was his total fluid intake?

    2.95 liters

17. The distance between toll booths on a certain stretch of highway is 29.375 miles. Within this distance highway construction is taking place for 5.8 miles and 3.75 miles. How many miles of highway are not under construction within this distance?

    19.825 miles

18. Karen's car holds 22.8 gallons of gas. She needed to purchase 11.37 gallons to fill the tank. How many gallons of gas were already in the tank?

    11.43 gallons

19. Four books are placed side by side on a shelf measuring 63.48 centimeters long. If the books' thicknesses are 4.5 centimeters, 4.375 centimeters, 8.42 centimeters, and 7.9 centimeters, how many centimeters of shelf space are left?

    38.285 centimeters

20. Hector's height increased from 49.5 inches to 50.25 inches during a certain period. Over the same time span, Lewis's height increased from 43.75 inches to 45 inches. By how many inches did Lewis's growth exceed Hector's?

    .5 inch

UNIT 36 Multiplication of Decimals

Multiplying decimal numbers is a straightforward procedure, because decimal points *do not* have to be lined up.

1 Arrange and multiply the decimal numbers as if they were whole numbers.

2 Count the total number of digits to the right of each decimal point in the multiplier and the multiplicand.

3 In the product, starting from the right, count to the left the number of decimal places equal to the number of digits counted in step 2. Insert the decimal point at this point.

4 If the total number of places to be counted is greater than the number of places in the product, insert as many zeros as necessary at the left of the product.

EXAMPLE 1 **Multiply: 6.24 × 1.3.**

$$\begin{array}{r} 6.24 \\ \times \quad 1.3 \\ \hline 1.872 \\ 624 \\ \hline 8.112 \end{array}$$

1. Arrange and multiply.
2. There are 3 digits to the right of each decimal point.
3. Count 3 places from right to left.

Answer: 8.112

PRACTICE 1 **Multiply: 4.54 × 2.6.**

Answer: 11.804

EXAMPLE 2 **Multiply: .032 × 1.4.**

$$\begin{array}{r} .032 \\ \times \quad 1.4 \\ \hline 128 \\ 032 \\ \hline .0448 \end{array}$$

1. Arrange and multiply.
2. There are 4 digits to the right of each decimal point.
3. Count 4 places from right to left. Add 1 zero at the left of the product.

Answer: .0448

PRACTICE 2 **Multiply: .0319 × .051.**

Answer: .0016269

UNIT 36 Name ______________________ Date ____________

Multiplication of Decimals

Multiply:

1. 56.2 × 4.3 =

   241.66

2. 21.5 × .27 =

   5.805

3. 7.1 × 8.2 =

   58.22

4. 8.123 × .09 =

   .73107

5. 4.5 × .15 =

   .675

6. 450 × .02 =

   9

7. 341.45 × .007 =

   2.39015

8. .132 × .241 =

   .031812

9. .23 × .009 =

   .00207

10. 7.02 × 5.27 =

    36.9954

11. 2.461 × .0529 =

    .1301869

12. 25.238 × 12.1 =

    305.3798

13. .132 × .241 =

    .031812

14. .01346 × 2.06 =

    .0277276

15. 94.263 × 5.57 =

    525.04491

16. 74.1 × 82 =

    6,076.2

17. 8.079 × .423 =

    3.417417

18. 72.9 × .0301 =

    2.19429

19. 81.5 × .0142 =

    1.1573

20. .347 × .026 =

    .009022

21. .4127 × 1.001 =

    .4131127

22. .3413 × 3.008 =

    1.0266304

23. 25.920 × 6.4 =

    165.888

24. 94.094 × 1.43 =

    134.55442

Business Applications:

25. A mortgage company reimburses its employees at the rate of 0.33 cents per mile. Assume a travel voucher has been submitted documenting mileage of 750.8 miles. Multiply these numbers to find the amount to be reimbursed. (Round the answer to the nearest cent.)

    $247.76

26. A computer company wishes to rent 112,500 square feet of space at the cost of $3.75 per square foot. Multiply these numbers to find the total cost.

    $421,875.00

UNIT

37 Division of Decimals

When dividing decimal numbers by a whole number:

1. Place the decimal point in the quotient directly above the decimal point in the dividend.
2. Divide as in whole numbers.
3. If the division does not end, add as many zeros as necessary to make the quotient come out even, or to be able to round the answer to a given place value.
4. If necessary, round the answer to the specific place value by carrying the division out to one place beyond the specified place value.

When dividing a decimal number by another decimal number, we move the decimal points in the divisor and the dividend an equal number of places. Doing this does not affect the value of the quotient, because we've essentially multiplied the divisor and the dividend by the same number.

When dividing a decimal by a decimal:

1. Move the decimal point in the divisor to the right of the last digit, making it a whole number.
2. Move the decimal point in the dividend an equal number of places to the right as the decimal point in the divisor was moved. Insert zeros in the dividend if necessary.
3. Place the decimal point in the quotient directly above the decimal point in the dividend.
4. Divide as in whole numbers.
5. If the division does not end, continue dividing by adding as many zeros as necessary to make the quotient come out even or to be able to round the answer to a given place.
6. If necessary, round the answer.

EXAMPLE 1 **Divide the following.**

84.72 ÷ 6 =

151.32 ÷ 13 =

Answers:

```
   14.12
6 )84.72
```

```
     11.64
13 )151.32
    13
     21
     13
      83
      78
       52
       52
        0
```

PRACTICE 1 **Divide the following.**

2.928 ÷ 4 = 632.06 ÷ 26 =

Answers: .732 24.31

EXAMPLE 2 **Divide: .32085 ÷ .15.**

.15) .32085

1. Move decimal point in the divisor and the dividend 2 places to the right.
2. Divide as in whole numbers.

Answer:

```
      2.139
15 ) 32.085
     30
      20
      15
       58
       45
       135
       135
         0
```

PRACTICE 2 **Divide: 9.6968 ÷ .23.**

Answer: 42.16

EXAMPLE 3 **Divide: 16 ÷ .003 (round to the nearest tenth if necessary).**

.003) 16

1. Move decimal point in the divisor and the dividend three places to the right, adding 3 zeros to the dividend.

```
   5,333.33
3 ) 16,000.00
```

2. Divide as in whole numbers, adding zeros to be able to round to the nearest tenth.

Answer: 5,333.33 rounded = 5,333.3

PRACTICE 3 **Divide: 96.47 ÷ 3.12 (round to the nearest hundredth if necessary).**

Answer: 30.92

UNIT **37** Name ______________________ Date ______________

Division of Decimals

Divide and round the answer to the specified place value when stated:

1. 1.032 ÷ 6 =
   .172
2. 750.6 ÷ 9 =
   83.4
3. 602.16 ÷ 13 =
   46.32
4. .72 ÷ 24 =
   .03
5. .0126 ÷ 7 =
   .0018
6. .285 ÷ 34; hundredth =
   .01
7. 513.24 ÷ 52 =
   9.87
8. .628 ÷ 14; thousandth =
   .045
9. 68.851 ÷ 27; tenth =
   2.6
10. 84.273 ÷ 21 =
    4.013
11. 20.15 ÷ 8; thousandth =
    2.519
12. .7 ÷ .2 =
    3.5
13. .284 ÷ .04 =
    7.1
14. 468.16 ÷ 7.7 =
    60.8
15. .6322 ÷ 5.8 =
    .109
16. 1739 ÷ 3.7 =
    470
17. 97.29 ÷ .047 =
    2,070
18. 220.16 ÷ .0128 =
    17,200
19. 94.094 ÷ .143 =
    658
20. 70 ÷ 13.2; hundredth =
    5.30
21. 13.3 ÷ 8.4; hundredth =
    1.58
22. 70.7 ÷ .87; tenth =
    81.3
23. .776 ÷ 69; thousandth =
    .011
24. 7.66 ÷ 2.6; thousandth =
    2.946
25. .001211 ÷ .173 =
    .007
26. 3.9256 ÷ .037; tenth =
    106.1
27. 12.119 ÷ .005 =
    2,423.8
28. .8325 ÷ 2.6; ten thousandth =
    .3202

Business Applications:

29. A postal service must distribute 1,875.465 pounds of mail equally among 3 vehicles. Divide to find the weight to be placed in each vehicle.

625.155 pounds

30. A real estate sales office's heating bill for 25.5 days was $378.40. Divide these numbers to find the average daily heating cost. Round the answer to the nearest dollar.

$15

UNIT

38 Multiplication and Division of Decimals by Multiples of Ten

When multiplying or dividing decimals by numbers that are multiples of 10, it is useful to take advantage of the following rules:

Multiplication:

1. Count the end zeros in the multiplier.
2. Move the decimal point in the multiplicand the same number of places *to the right* as there are end zeros in the multiplier.
3. Insert zeros to the right of the multiplicand, if necessary.

Division:

1. Count the end zeros in the divisor.
2. Move the decimal point in the dividend the same number of places *to the left* as there are end zeros in the divisor.
3. Insert zeros to the left of the dividend, if necessary.

EXAMPLE 1 **Multiply the following.**

Answers:

$4.35 \times 10 \Rightarrow$ *one* place to right $=$ 43.5

$4.35 \times 100 \Rightarrow$ *two* places to right $=$ 435

$4.35 \times 1{,}000 \Rightarrow$ *three* places to right so a zero must be inserted $=$ 4,350

PRACTICE 1 **Multiply the following.**

Answers: $13.28 \times 10 =$ 132.8 $\quad 13.28 \times 100 =$ 1,328 $\quad 13.28 \times 1{,}000 =$ 13,280

EXAMPLE 2 **Divide the following.**

Answers:

$27.6 \div 10 \Rightarrow$ *one* place to left $=$ 2.76

$27.6 \div 100 \Rightarrow$ *two* places to left $=$ .276

$27.6 \div 1{,}000 \Rightarrow$ *three* places to left so a zero must be inserted $=$ .0276

PRACTICE 2 **Divide the following.**

Answers: $89.1 \div 10 =$ 8.91 $\quad 89.1 \div 100 =$ .891 $\quad 89.1 \div 1{,}000 =$ .0891

UNIT 38 Name ______ Date ______

Multiplication and Division of Decimals by Multiples of Ten

Multiply:

1. 7.38 × 10 = 73.8
2. 9.21 × 100 = 921
3. 4.72 × 1,000 = 4,720
4. 6.3 × 100 = 630
5. 3.27 × 10 = 32.7
6. .596 × 10,000 = 5,960
7. .987 × 100 = 98.7
8. 3.247 × 1,000 = 3,247
9. 5.63 × 10 = 56.3
10. 4.2 × 100 = 420
11. .085 × 100 = 8.5
12. .071 × 10 = .71
13. 4.98 × 10,000 = 49,800
14. .021 × 1,000 = 21
15. 3.3 × 1,000 = 3,300
16. 6.96 × 100 = 696
17. .508 × 10 = 5.08
18. .0469 × 10,000 = 469
19. 9.76 × 10 = 97.6
20. 27.1 × 100 = 2,710

Divide:

21. .843 ÷ 10 = .0843
22. 9.79 ÷ 100 = .0979
23. 33.176 ÷ 100 = .33176
24. .74 ÷ 1,000 = .00074
25. 5.63 ÷ 1,000 = .00563
26. 9.16 ÷ 10 = .916
27. 5.9 ÷ 100 = .059
28. 73.8 ÷ 10 = 7.38
29. 9.2 ÷ 100 = .092
30. .20 ÷ 1,000 = .0002
31. 6.47 ÷ 100 = .0647
32. 38.96 ÷ 10,000 = .003896
33. 739.1 ÷ 1,000 = .7391
34. .038 ÷ 10 = .0038
35. 47.6 ÷ 10,000 = .00476
36. 1.190 ÷ 1,000 = .00119
37. 3.318 ÷ 10 = .3318
38. 618.9 ÷ 100 = 6.189
39. 27.94 ÷ 10,000 = .002794
40. 13.99 ÷ 1,000 = .01399

UNIT **39**

How to Dissect and Solve Word Problems: Multiplication and Division of Decimals

Use the chart below as an aid to solving the word problems in this unit.

Gather the facts.	What am I solving for?	What must I need to know or calculate before solving the problem?	Key points to remember.

EXAMPLE 1 **Nancy earns $18.75 per hour. If she worked 37.5 hours last week, how much did she earn (round your answer to the nearest hundredth)?**

Gather the facts.	What am I solving for?	What must I need to know or calculate before solving the problem?	Key points to remember.
Hourly rate = $18.75 Hours worked = 37.5	Total earnings	Hourly rate × Hours worked = Total earnings	Count digits to right of decimal points. Place decimal point in answer by moving right to left. Round answer to hundredth.

```
   $18.75
 ×   37.5
 --------
     9375
   13125
   5625
 --------
```

Answer: $703.125 rounded = $703.13

PRACTICE 1 **Jumbo size shrimp cost $11.98 per pound. What is the total cost of 3.5 pounds of shrimp?**

Answer: $41.93

EXAMPLE 2 **A distance of 225.5 miles is traveled at a constant rate of 65 miles per hour. How long does it take to travel this distance (round your answer to nearest hundredth)?**

Gather the facts.	What am I solving for?	What must I need to know or calculate before solving the problem?	Key points to remember.
Distance = 225.5 miles Rate = 65 miles per hour	Time spent traveling	Distance ÷ Rate = Time spent traveling	Place decimal point in quotient directly above decimal point in dividend. Round to nearest hundredth.

Answer:

```
      3.469   rounded = 3.47
65 ) 225.500
     195
      305
      260
       450
       390
        600
        585
         15
```

PRACTICE 2 **Maryanne drove a distance of 73.5 miles and used 3.5 gallons of gas. How many miles per gallon of gas did she get?**

Answer: 21 miles per gallon

UNIT **39** Name ______________________ Date ____________

How to Dissect and Solve Word Problems: Multiplication and Division of Decimals

Solve the following problems:

1. Andre wishes to build a bookcase having 5 shelves. The length of each shelf is to be 52.75 inches. What is the total length in inches of the 5 shelves (rounded to the nearest tenth)?

   263.8 inches

2. A college football running back averages 4.7 yards each time he carried the ball. He carried the ball 23 times. What was his total yardage?

   108.1 yards

3. A stack of equal-sized patio blocks measures 28.5 inches high. Each block is 2.375 inches thick. How many blocks are in the stack?

   12 blocks

4. Paula runs a distance of 11.2 miles in 72.8 minutes. How many minutes per mile does she run?

   6.5 minutes

5. Ana is paying $328.09 per month to pay off a car loan of $6,233.71. How many months will it take to pay off the loan?

   19 months

6. What is the cost of a 5.25 pound pork loin roast if it costs $4.39 per pound? (Round to the nearest cent.)

   $23.05

7. The net weight of a can of coffee is 11.5 ounces. One can of coffee costs $1.99. How much per ounce does the coffee cost? (Round to the nearest cent.)

   $ .17

8. How much will 14.25 yards of fabric cost if the cost per yard is $13.50 (round to the nearest cent)?

   $192.38

9. An average of 34.4 people stop for gas at a self-service station each hour. The station is open 20.5 hours each day. How many people stop there for gas? (Round to the nearest whole number.)

   705 people

10. How many centimeters are there in 11.623 inches if 1 centimeter equals .394 inches?

    29.5 centimeters

11. What is the total amount to be paid for a computer system if Don is making 19 payments of $93.67?

$1,779.73

12. Two 8.5-foot lengths of wood cost $13.43. What is the cost per linear foot?

.79 cents per linear foot

13. Caitlin's grade point summaries for 3 terms were 3.125, 3.35, and 3.05. What is her average grade point summary?

3.175

14. Find the total cost of a 3.5-pound roast at $3.79 per pound and 10 pounds of potatoes at $.39 per pound. (Round to the nearest cent.)

$17.17

15. How long will a 250-cubic-centimeters bag of dextrose and water last if it drips at a rate of 30 cubic centimeters per hour (round to the nearest hour)?

8 hours

16. If 1,000 cubic centimeters equals 1 liter, how many cubic centimeters are there in 12.79 liters?

12,790 cubic centimeters

17. A 250 milligram medication is prescribed 3 times a day. What is the total intake of milligrams over 7.5 days?

5,625 milligrams

18. A kilowatt hour of electricity costs $35.77. What is Willie's electric bill if he used 2.43 kilowatt hours of electricity? (Round to the nearest cent.)

$86.92

19. Coal costs $176.50 for the first 2,000 pounds purchased and $3.19 per 50 pounds thereafter. Find the total cost of 3,500 pounds of coal?

$272.20

20. A 3-pound bag of lettuce mixture costs $2.99 at Food Warehouse X. Two 3-pound bags of the same lettuce mixture cost $3.49 at Food Warehouse Y. How much money per 3-pound bag does one save purchasing the mixture at Food Warehouse Y? (Round to the nearest cent.)

$1.25 per 3-pound bag

UNIT

40 Converting Fractions to Decimals

Every fraction can be rewritten as a decimal because the line separating the numerator and denominator of a fraction means to divide. We use this fact to rewrite fractions as decimals. Remember that the decimal point is not shown in a whole number. It is understood to be located at the very end of the number.

Converting any fraction to a decimal can be done by using the following procedures:

1. Examine the denominator. If it is a multiple of ten (10, 100, 1,000, 10,000, etc.), move the decimal point in the numerator (understood or otherwise) to the left as many places as there are end zeros in the denominator.
2. If the denominator is not a multiple of ten, divide the numerator by the denominator. Add zeros to the dividend as necessary and continue dividing until the division stops.
3. If the division does not stop, round to the appropriate place value.
4. When converting a mixed number to a decimal, add the whole number part to the converted fraction part.

EXAMPLE 1 **Convert to decimals.**

Answers:

$\frac{3}{10}$ ⇒ *one* zero = .3

$\frac{36}{100}$ ⇒ *two* zeros = .36

$\frac{3.6}{1,000}$ ⇒ *three* zeros = .0036

$\frac{36}{10,000}$ ⇒ *four* zeros = .0036

PRACTICE 1 **Convert to decimals.**

Answers: $\frac{7}{100}$ = .07 $\quad \frac{14}{1,000}$ = .014 $\quad \frac{7.2}{1,000}$ = .0072 $\quad \frac{72}{1,000}$ = .072

EXAMPLE 2 **Change to a decimal (round to the nearest thousandth when necessary).**

Answers: a. $\frac{3}{5} = 5\overline{)3.0}^{\,.6} = .6$ b. $\frac{1}{12} = 12\overline{)1.0000}^{\,.0833} = .083$

$$\begin{array}{r} .0833 \\ 12\overline{)1.0000} \\ 00 \\ \hline 100 \\ 96 \\ \hline 40 \\ 36 \\ \hline 4 \end{array}$$

PRACTICE 2 **Change to a decimal (round to the nearest thousandth when necessary).**

Answers: $\frac{2}{25}$ = .08 $\frac{1}{9}$ = .111

EXAMPLE 3 **Change $4\frac{2}{5}$ to a decimal.**

Answer: $4\frac{2}{5} = 4 + 5\overline{)2.0}^{\,.4} = 4 + .4 = 4.4$

PRACTICE 3 **Change $16\frac{1}{4}$ to a decimal:**

Answer: $16\frac{1}{4}$ = 16.25

UNIT 40 Name ______ Date ______

Converting Fractions to Decimals

Convert the following fractions to decimals (round to the nearest thousandth if necessary).

1. $\frac{16}{100}$ = .16
2. $\frac{22}{1,000}$ = .022
3. $4\frac{3}{100}$ = 4.03
4. $\frac{14}{10,000}$ = .0014
5. $\frac{7.7}{100}$ = .077
6. $\frac{32.5}{1,000}$ = .0325
7. $\frac{2}{9}$ = .222
8. $6\frac{1}{2}$ = 6.5

UNIT **40** Name ______________________ Date ______________

Converting Fractions to Decimals

9. $2\frac{5}{8}$ = 2.625

10. $\frac{3}{14}$ = .214

11. $\frac{4}{9}$ = .444

12. $\frac{1}{4}$ = .25

13. $\frac{1}{5}$ = .2

14. $\frac{5}{6}$ = .833

15. $\frac{23}{24}$ = .958

16. $4\frac{1}{5}$ = 4.2

17. $\frac{48}{100}$ = .48

18. $\frac{1}{7}$ = .143

19. $\frac{2}{3}$ = .667

20. $11\frac{3}{8}$ = 11.375

21. $\frac{4.75}{10{,}000}$ = .000475

22. $\frac{1}{2}$ = .5

23. $\frac{53}{1{,}000}$ = .053

24. $\frac{4.4}{5}$ = .88

25. $\frac{11}{34}$ = .324

26. $\frac{3}{7}$ = .429

27. $\frac{17}{21}$ = .810

28. $\frac{3.2}{16}$ = .2

29. $\frac{1}{6}$ = .167

30. $5\frac{7}{9}$ = .778

31. $\frac{15}{4}$ = 3.75

32. $\frac{140}{1{,}000}$ = .14

33. $\frac{7}{8}$ = .875

34. $\frac{35}{24}$ = 1.458

35. $\frac{16}{17}$ = .941

36. $8\frac{2}{5}$ = 8.4

37. $\frac{1.7}{23}$ = .074

38. $\frac{5}{1{,}000}$ = .005

Business Applications:

39. The closing price of Polaris stock was listed as $20\frac{3}{4}$. Express this as a decimal number.

20.75

40. The net change for Pineapple Graphics stock was $+\frac{3}{8}$. Express this as a decimal number.

+ .375

Converting Decimals to Fractions

Converting decimals to fractions can be done simply by writing what one hears oneself say. Thus, it is important to be able to read decimal numbers accurately.

Follow these rules:

1 Convert the decimal to a fraction by writing exactly what you say. Keep in mind that:

a. "Tenths" means a denominator of 10.

b. "Hundredths" means a denominator of 100.

c. "Thousandths" means a denominator of 1,000.

d. "Ten thousandths" means a denominator of 10,000.

2 Reduce the fraction to lowest terms.

3 Mixed numbers will be formed when a whole number exists to the left of the decimal point.

A special situation exists when the decimal to be converted is actually a mixed decimal number. This type of a decimal consists of a decimal part and a fraction part. An example of this is $.48\frac{1}{2}$ where the decimal part is .48 and the fraction part is $\frac{1}{2}$. Converting this kind of decimal is done in this manner.

1 Disregard the decimal point and rewrite the mixed decimal as an improper fraction.

2 Divide this improper fraction by the number that is the place value of the last digit of the decimal part.

3 Reduce the fraction to lowest terms if necessary.

EXAMPLE 1 **Rewrite as fractions.**

Answers:

$.013 \Rightarrow$ thirteen thousandths $= \frac{13}{1{,}000}$

$.44 \Rightarrow$ forty-four hundredths $= \frac{44}{100} = \frac{11}{25}$

$16.75 \Rightarrow$ sixteen and seventy-five hundredths $= 16\frac{75}{100} = 16\frac{3}{4}$

PRACTICE 1 **Rewrite as fractions.**

Answers: $.22 = \frac{11}{50}$ $.007 = \frac{7}{1{,}000}$ $33.16 = 33\frac{4}{25}$

EXAMPLE 2 **Rewrite $.08\frac{3}{5}$ as a fraction.**

$$8\frac{3}{5} = \frac{43}{5}$$

1. Disregard the decimal point and write $8\frac{3}{5}$ as an improper fraction.

Answer: $$\frac{\frac{43}{5}}{100} = \frac{43}{5} \div \frac{100}{1} = \frac{43}{5} \times \frac{1}{100} = \frac{43}{500}$$

2. Divide $\frac{43}{5}$ by 100 because 8 sits in the hundredths position in the original mixed decimal.

PRACTICE 2 **Rewrite $.16\frac{1}{4}$ as a fraction.**

Answer: $\frac{13}{80}$

UNIT 41 Name ______________________ Date ____________

Converting Decimals to Fractions

Convert the following decimals to fractions:

1. $.09 = \frac{9}{100}$
2. $.101 = \frac{101}{1,000}$
3. $.36 = \frac{9}{25}$
4. $.04 = \frac{1}{25}$
5. $.176 = \frac{22}{125}$
6. $.37 = \frac{37}{100}$
7. $.375 = \frac{3}{8}$
8. $.13 = \frac{13}{100}$
9. $.178 = \frac{89}{500}$
10. $.3367 = \frac{3,367}{10,000}$
11. $.22 = \frac{11}{50}$
12. $4.32 = 4\frac{8}{25}$
13. $17.22 = 17\frac{11}{50}$
14. $.625 = \frac{5}{8}$
15. $.024 = \frac{3}{125}$
16. $4.314 = 4\frac{157}{500}$

UNIT **41** Name ______________ Date ______________

Converting Decimals to Fractions

17. $36.42 = 36\frac{21}{50}$

18. $105.7 = 105\frac{7}{10}$

19. $30.84 = 30\frac{21}{25}$

20. $.00481 = \frac{481}{100{,}000}$

21. $.000006 = \frac{3}{500{,}000}$

22. $.000014 = \frac{7}{500{,}000}$

23. $12.8 = 12\frac{4}{5}$

24. $68.1875 = 68\frac{3}{16}$

25. $10.44 = 10\frac{11}{25}$

26. $22.035 = 22\frac{7}{200}$

27. $7.584 = 7\frac{73}{125}$

28. $5.625 = 5\frac{5}{8}$

29. $.216 = \frac{27}{125}$

30. $.16\frac{3}{4} = \frac{67}{400}$

31. $.43\frac{3}{4} = \frac{7}{16}$

32. $.008\frac{1}{5} = \frac{41}{5{,}000}$

33. $.06\frac{1}{5} = \frac{31}{500}$

Business Applications:

34. The Conference Board's index of consumer confidence shows a decline of 6.75 points. Express this number as a reduced fraction.

$6\frac{3}{4}$ points

35. A national health care consultant to a community hospital reports to the president that the RN staff spends on average 3.6 hours per day on non-nursing care tasks. Express this number as a reduced fraction.

$3\frac{3}{5}$ hours

UNIT

42 How to Dissect and Solve Word Problems: Banking and Credit Card Transactions

An often-seen application of the use of decimals is in banking. The activities of calculating net deposits for credit card transactions and reconciling bank statements with checkbook balances are just two banking activities where use of decimal numbers occurs.

When finding the net deposit for credit card transactions:

1. Add all credit card sales amounts.
2. Add all credit card return amounts.
3. Subtract the total returns from the total sales to find the net deposit.

EXAMPLE 1 **Find the net deposit after the following transactions took place.**

Credit card sales: $17.08, $132.99, $46.99, $89.99, $62.57
Returns: $9.08, $27.19, $53.99

Gather the facts.	What am I solving for?	What must I need to know or calculate before solving the problem?	Key points to remember.
Sales: $17.08 $132.99 $46.99 $89.99 $62.57 Returns: $9.08 $27.19 $53.99	Net deposit	Credit card sales − Returns = Net deposit	Line up decimal points when adding and subtracting.

Sales:

$17.08
$132.99
$46.99
$89.99
\+ $62.57
$349.62

Returns:

$9.08
$27.19
\+ $53.99
$90.26

Net deposit:

$349.62
− $90.26

Answer: $259.36

PRACTICE 1 **Find the net deposit given the following transactions.**

Credit card sales:	\$129.89, \$36.36, \$141.23, \$79.99, \$11.08
Returns:	\$12.15, \$34.01, \$22.83

Answer: \$329.56

To perform a bank reconciliation:

1 Add to the *bank balance* all outstanding deposits (known deposits not yet processed by the bank).

2 Subtract from the sum obtained in step 1 all outstanding checks (checks written but not yet processed by the bank) to find the *final bank balance.*

3 Add to the checkbook balance any interest payments or other deposits made by the bank to the checking account.

4 Subtract from the sum obtained in step 3 any fees or deductions made by the bank to the checking account. The amount obtained is the *final checkbook balance.*

5 The final bank and checkbook balances should agree.

Figures 3.42.1 and 3.42.2 are examples of a bank statement and a bank reconciliation that was completed using the information from the bank statement. The notations "DM" on the bank statement are debit memos or deductions made by the bank to the checking account. The notation "CM" is a credit memo or addition made by the bank to the checking account.

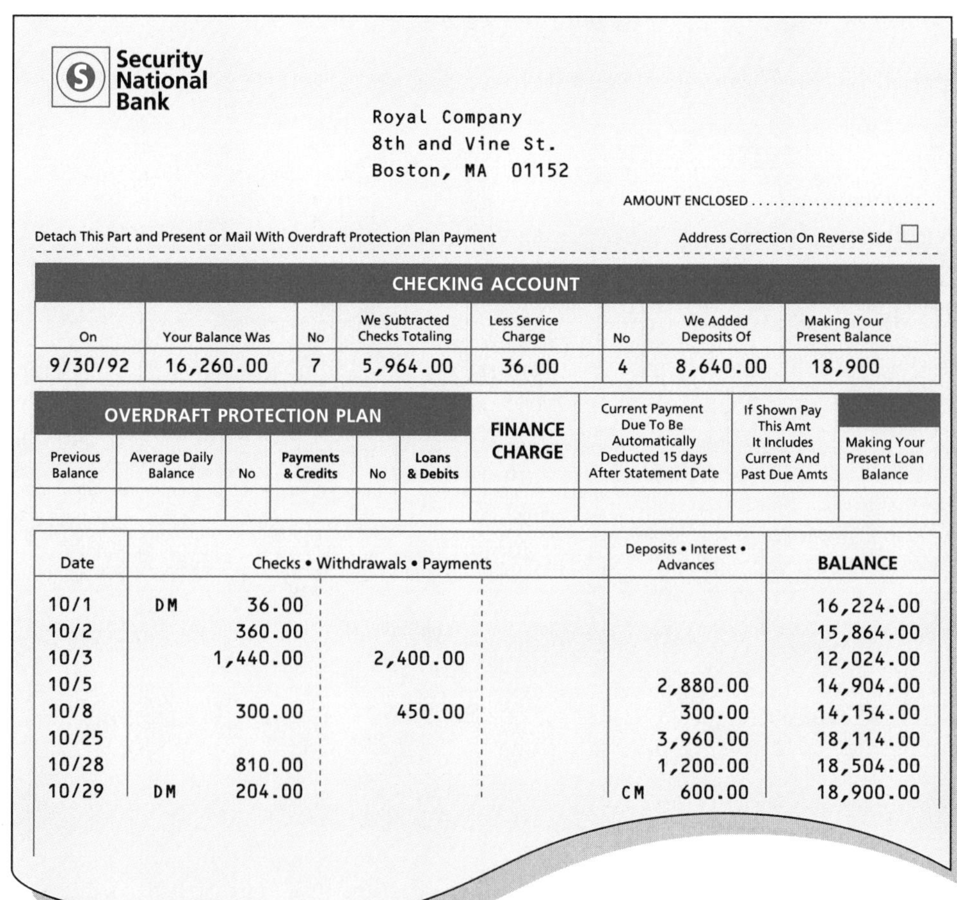

Security National Bank

Royal Company
8th and Vine St.
Boston, MA 01152

AMOUNT ENCLOSED ....

Detach This Part and Present or Mail With Overdraft Protection Plan Payment — Address Correction On Reverse Side ☐

CHECKING ACCOUNT

On	Your Balance Was	No	We Subtracted Checks Totaling	Less Service Charge	No	We Added Deposits Of	Making Your Present Balance
9/30/92	16,260.00	7	5,964.00	36.00	4	8,640.00	18,900

OVERDRAFT PROTECTION PLAN						FINANCE CHARGE	Current Payment Due To Be Automatically Deducted 15 days After Statement Date	If Shown Pay This Amt It Includes Current And Past Due Amts	Making Your Present Loan Balance
Previous Balance	Average Daily Balance	No	Payments & Credits	No	Loans & Debits				

Date	Checks • Withdrawals • Payments			Deposits • Interest • Advances	BALANCE
10/1	DM	36.00			16,224.00
10/2		360.00			15,864.00
10/3		1,440.00	2,400.00		12,024.00
10/5				2,880.00	14,904.00
10/8		300.00	450.00	300.00	14,154.00
10/25				3,960.00	18,114.00
10/28		810.00		1,200.00	18,504.00
10/29	DM	204.00		CM 600.00	18,900.00

Figure 3.42.1
Bank Statement

Figure 3.42.2
Bank Reconciliation

ROYAL COMPANY
Bank Reconciliation as of October 31, 1992

Checkbook Balance			Bank Balance		
Checkbook balance of cash		$18,690	Bank statement balance		$18,900
Add:			Add:		
Collection of note		600	Deposits in transit 10/29		1,000
		$19,290			$19,900
Deduct:			Deduct:		
Check printing	$ 36		Outstanding checks:		
			No. 108	$180	
NSF	204	240	No. 110	670	850
Reconciled balance		$19,050	Reconciled balance		$19,050

EXAMPLE 2

Matthew's checkbook balance as of February 28 was $1,623.63. His bank statement shows a balance of $1,862.28. An outstanding check for $637.29 is listed along with an outstanding deposit of $400.00. The bank statement shows a deposit of $7.86 for earned interest and a fee of $6.50 for new checks. Prepare a bank reconciliation.

Gather the facts.	What am I solving for?	What must I need to know or calculate before solving the problem?	Key points to remember.
Checkbook balance: $1,623.63 Bank balance: $1,862.28 Outstanding checks: $637.29 Outstanding deposits: $400.00 Fees: $6.50 Interest: $7.86	Reconciled balance	Beginning balance + Outstanding deposits − Outstanding checks = Bank balance Beginning checkbook balance + Deposits − Fees = Checkbook balance	Line up decimal points.

	Bank Balance		Checkbook Balance	
	Bank statement balance	$1,862.28	Checkbook balance	$1,623.63
	Add:		*Add:*	
	Outstanding deposit	+ $400.00	Bank interest	+ $7.86
		$2,262.28		$1,631.49
	Subtract:		*Subtract:*	
	Outstanding check	− $637.29	New checks fee	− $6.50
Answer:	Final bank balance	$1,624.99	Final checkbook balance	$1,624.99

PRACTICE 2 **Prepare a bank reconciliation.**

Bank statement balance:	$5,981.50
Checkbook balance:	$5,790.40
Outstanding checks:	$290.00
	$160.00
Outstanding deposits:	$221.00
Service charge:	$50.00
Earned interest:	$12.10

Answer: $5,752.50

UNIT Name ______________________ Date ____________

42 *How to Dissect and Solve Word Problems: Banking and Credit Card Transactions*

Solve:

1. Credit card transactions on May 16 at Lenox Co. were:
   Sales: $318.59, $29.50, $196.37
   Returns: $52.16, $138.00
   Calculate the total net deposits.

   $354.30

2. Jordan's bank statement shows a balance of $922.16. His checkbook indicates a balance of $866.71. Outstanding checks total $58.26. The bank charged $7 for servicing and deposited $4.19 in earned interest. Prepare a bank reconciliation.

   $863.90

3. Harris's Men's Wear had the following credit card transactions:
   Sales: $24.99, $19.80, $63.66, $144.80, $28.75, $46.19
   Returns: $38.01, $16.50, $37.99
   Calculate the total net deposit.

   $235.69

4. Julie's checkbook balance is $1,071.37. Her bank statement shows a balance of $965.18, a deposit of $6.36 in earned interest, and charges totaling $11.50. Outstanding checks total $248.95, and there is an outstanding deposit of $350. Prepare a bank reconciliation.

   $1,066.23

UNIT 42 Name ______________________ Date ______________

How to Dissect and Solve Word Problems: Banking and Credit Card Transactions

5. Prepare a bank reconciliation given the following information:

Outstanding checks:	$2,940.35
Outstanding deposits:	$3,626.80
Earned interest:	$13.70
Bank charges:	$229.00
Bank statement balance:	$9,740.35
Checkbook balance:	$10,642.10

$10,426.80

6. Yardley Comics notes that on December 17 its credit card sales amounted to $329.08. It has slips for credit card returns for $17.22 and $14.39. Find the total net deposit.

$297.47

7. Rick's credit card purchases this past Christmas were $69.37, $119.99, $26.23, $11.75, and $249.99. Credit card returns totaled $43.45. What is the net purchase amount?

$433.88

8. Maria's checkbook reveals she has a balance of $7,041.23. Her bank statement shows a balance of $8,119.21. The bank statement also indicates earned interest in the amount of $21.32 and charges totaling $42. An outstanding deposit amounts to $750.00, and outstanding checks are for $229.99, $1,017.00, and $601.67. Prepare a bank reconciliation.

$7,020.55

9. Abington Pet Supplies received its bank statement, and it showed a balance of $9,400.81. Their checkbook balance is $8,700.85. Outstanding checks total $1,223.61, and there is an outstanding deposit of $500. The bank statement shows a $32.15 checking fee and a deposit for $8.50 earned interest. Prepare a bank reconciliation.

$8,677.20

10. José's checkbook balance shows a total of $96.77. His bank statement reveals a balance of $1,142.36, along with a deposit of $2.19 earned interest, and a $6 servicing fee. He has outstanding checks recorded for $750.00, $49.40, $63.87, and $186.13. Prepare his bank reconciliation.

$92.96

11. Walton's Savings and Trust sent Alexis her bank statement. Its contents were:

Balance:	$1,571.23
Charges:	$32.19
New checks fee:	$10.39
Earned interest:	$8.62

Alexis's checkbook balance is $1,609.12. She has recorded an outstanding deposit for $229 and outstanding checks totaling $225.07. Prepare a bank reconciliation.

$1,575.16

12. Randy's checking account has a balance of $1,765.69. His outstanding checks total $1,072.16, and his outstanding deposits total $1,632.19. A bank charge of $6 appears on his statement, along with a deposit of $12.19 in interest. Prepare a bank reconciliation for him if the bank statement shows a balance of $1,211.85.

$1,771.88

13. Mercedes's checkbook balance is $37.08. Her bank statement indicated a balance of $1,232.66, along with charges totaling $18.00, and earned interest amounting to $2.37. She has recorded checks for $22.37, $48.16, $66.99, $43.56, $118.91, $32.67, and $878.55 that are still outstanding. Prepare a bank reconciliation.

$21.45

14. World of Trains has a record of credit card sales and returns for Saturday, February 17, that look like:

Sales:	Returns:
$17.23	$12.16
$144.19	$7.08
$39.99	
$22.08	
$67.41	
$56.59	

Calculate their total net deposit.

$328.25

15. Hank's bank statement shows a balance of $2,229.00, and earned interest deposit of $7.32, and a deduction of $12.50 for new checks. An outstanding deposit exists for $400.00, and outstanding checks total $842.52. Prepare a bank reconciliation, assuming his checkbook has a balance of $1,791.66.

$1,786.48

Preparing a Payroll

Another practical application of the use of decimals occurs in the preparation of payrolls. Certainly the number of deductions made in preparing a typical payroll are many more than those cited in the following exercises. Nevertheless, completing the "simplified" payroll exercises that follow gives one an idea of the role decimal numbers play in the process.

To prepare a payroll:

1. Add the hours worked by the employee.
2. Multiply the sum obtained in step 1 by the hourly rate to obtain the gross wages.
3. Determine the appropriate amount of federal withholding tax from Table 3.43.1, which appears on page 128.
4. Multiply the gross wages amount by .062 to determine the FICA–Social Security deduction.
5. Multiply the gross wage amount by .0145 to determine the FICA–Medicare tax deduction.
6. Add the federal withholding tax, the Social Security tax, and the Medicare tax to find the total amount to be deducted from the gross wage.
7. Subtract the total obtained in step 6 from the gross wages to obtain the net pay.

EXAMPLE 1 **Determine the gross wages for the following individual (round gross wages to the nearest cent).**

Employee Name	Hours Worked	Total Hours	Hourly Rate	Gross Wages
Dobson, Elaine	S M T W Th F S 0 8 8 10 7 8 6.5	47.5	\$8.65	\$410.88

8 + 8 + 10 + 7 + 8 + 6.5 = 47.5
\$8.65 × 47.5 = \$410.875

Answer: \$410.875 rounded to the nearest cent = \$410.88

PRACTICE 1 **Determine the gross wages for the following individual (round gross wages to the nearest cent).**

Employee Name	Hours Worked	Total Hours	Hourly Rate	Gross Wages
Carson, Jeffrey	S M T W Th F S 5 8 5.5 8 8 8 2	44.5	\$6.50	\$289.25

Answer: \$289.25

EXAMPLE 2 **Using the table provided, calculate the net pay for the following individual (round to the nearest cent).**

Name	Withholding Allowance and Marital Status	Hours Worked	Hourly Rate	Gross Wages	Deductions				Net Pay
					Federal Withholding Tax	Social Security Tax	Medicare Tax	Total Deductions	
Sanders	S–2	37.5	$18.50	$693.75	$103.00	$43.01	$10.06	$156.07	$537.68

Gross wages = 37.5 × $18.50 = $693.75

Income tax from table 3.43.1 = $103.00

Social Security tax = .062 × $693.75 = $43.01 (rounded)

Medicare tax = .0145 × $693.75 = $10.06 (rounded)

Total deductions = $103.00 + $43.01 + $10.06 = $156.07

Answer: Net pay = $693.75 – $156.07 = $537.68

PRACTICE 2 **Using the table provided, calculate the net pay for the following individual (round to the nearest cent).**

Name	Withholding Allowance and Marital Status	Hours Worked	Hourly Rate	Gross Wages	Deductions				Net Pay
					Federal Withholding Tax	Social Security Tax	Medicare Tax	Total Deductions	
Barton	M–1	40	$22.35	$894.00	$118.00	$55.43	$12.96	$186.39	$707.61

Answer: $707.61

Name ______________________ Date ______________

Preparing a Payroll

Prepare the payrolls below:

Employee Name	Hours Worked S	M	T	W	Th	F	S	Total Hours	Hourly Rate	Gross Wage
Biller, Norman	0	8	8	8	7	3	4	38	$22.50	$855.00
Carter, Susan	0	8	8	8	7.5	5	4	40.5	$21.35	$864.68
Chidner, Jill	0	8	7.5	8	9	6	4	42.5	$22.50	$956.25
Dobbins, Edward	0	8	8	8	8	8	6.5	46.5	$23.75	$1,104.38
Dimitri, Sal	0	8	8	8	7.5	8	4	43.5	$22.50	$978.75
Fortin, Nelson	0	8	8	6	0	9.5	4	35.5	$21.35	$757.93
Flashberg, Amy	0	8	8	8	8	8	5.5	45.5	$21.35	$971.43
Harner, Stephen	4	7.5	8	8	8	0	0	35.5	$23.75	$843.13
Husson, Valerie	0	8	8	8	8	8.5	4.5	45	$22.50	$1.012.50
Koney, Stuart	5	8	8	7.5	8	9.5	0	46	$22.50	$1,035.00

Name	Withholding Allowance and Marital Status	Hours Worked	Hourly Rate	Gross Wages	Deductions				Net Pay
					Federal Withholding Tax	Social Security Tax	Medicare Tax	Total Deductions	
Atwell	S–0	37.5	$19.25	$721.88	$137.00	$44.76	$10.47	$192.23	$529.65
Autens	M–1	35.5	$21.25	$754.38	$89.00	$46.77	$10.94	$146.71	$607.67
Ayers	M–2	38	$21.25	$807.50	$89.00	$50.07	$11.71	$150.78	$656.72
Brown	S–0	39	$19.50	$760.50	$148.00	$47.15	$11.03	$206.18	$554.32
Buckley	S–0	40	$18.75	$750.00	$145.00	$46.50	$10.88	$202.38	$547.62
Green	S–1	37.5	$20.50	$768.75	$136.00	$47.66	$11.15	$194.81	$573.94
Harkins	M–2	40	$20.50	$820.00	$92.00	$50.84	$11.89	$154.73	$665.27
Hill	M–2	40	$21.25	$850.00	$97.00	$52.70	$12.33	$162.03	$687.97
Johnson	M–3	40	$21.25	$850.00	$90.00	$52.70	$12.33	$155.03	$694.97
Miller	M–1	38.5	$20.50	$789.25	$93.00	$48.93	$11.44	$153.37	$635.88

Table 3.43.1
Federal Withholding Tax Table

MARRIED Persons — **WEEKLY** Payroll Period

(For Wages Paid in 1993)

If the wages are —		And the number of withholding allowances claimed is —										
At least	But less than	0	1	2	3	4	5	6	7	8	9	10
		The amount of income tax to be withheld is —										
$740	$750	$94	$87	$80	$74	$67	$60	$53	$46	$40	$33	$26
750	760	95	89	82	75	68	61	55	48	41	34	28
760	770	97	90	83	77	70	63	56	49	43	36	29
770	780	98	92	85	78	71	64	58	51	44	37	31
780	790	100	93	86	80	73	66	59	52	46	39	32
790	800	103	95	88	81	74	67	61	54	47	40	34
800	810	106	96	89	83	76	69	62	55	49	42	35
810	820	108	98	91	84	77	70	64	57	50	43	37
820	830	111	99	92	86	79	72	65	58	52	45	38
830	840	114	101	94	87	80	73	67	60	53	46	40
840	850	117	104	95	89	82	75	68	61	55	48	41
850	860	120	107	97	90	83	76	70	63	56	49	43
860	870	122	110	98	92	85	78	71	64	58	51	44
870	880	125	113	100	93	86	79	73	66	59	52	46
880	890	128	115	103	95	88	81	74	67	61	54	47
890	900	131	118	106	96	89	82	76	69	62	55	49
900	910	134	121	108	98	91	84	77	70	64	57	50
910	920	136	124	111	99	92	85	79	72	65	58	52
920	930	139	127	114	101	94	87	80	73	67	60	53
930	940	142	129	117	104	95	88	82	75	68	61	55
940	950	145	132	120	107	97	90	83	76	70	63	56
950	960	148	135	122	110	98	91	85	78	71	64	58
960	970	150	138	125	112	100	93	86	79	73	66	59
970	980	153	141	128	115	103	94	88	81	74	67	61
980	990	156	143	131	118	105	96	89	82	76	69	62
990	1,000	159	146	134	121	108	97	91	84	77	70	64
1,000	1,010	162	149	136	124	111	99	92	85	79	72	65
1,010	1,020	164	152	139	126	114	101	94	87	80	73	67
1,020	1,030	167	155	142	129	117	104	95	88	82	75	68
1,030	1,040	170	157	145	132	119	107	97	90	83	76	70

SINGLE Persons — **WEEKLY** Payroll Period

(For Wages Paid in 1993)

If the wages are —		And the number of withholding allowances claimed is —										
At least	But less than	0	1	2	3	4	5	6	7	8	9	10
		The amount of income tax to be withheld is —										
$590	$600	$101	$88	$75	$63	$55	$48	$41	$35	$28	$21	$14
600	610	103	91	78	66	56	50	43	36	29	22	16
610	620	106	94	81	68	58	51	44	38	31	24	17
620	630	109	96	84	71	59	53	46	39	32	25	19
630	640	112	99	87	74	61	54	47	41	34	27	20
640	650	115	102	89	77	64	56	49	42	35	28	22
650	660	117	105	92	80	67	57	50	44	37	30	23
660	670	120	108	95	82	70	59	52	45	38	31	25
670	680	123	110	98	85	72	60	53	47	40	33	26
680	690	126	113	101	88	75	63	55	48	41	34	28
690	700	129	116	103	91	78	65	56	50	43	36	29
700	710	131	119	106	94	81	68	58	51	44	37	31
710	720	134	122	109	96	84	71	59	53	46	39	32
720	730	137	124	112	99	86	74	61	54	47	40	34
730	740	140	127	115	102	89	77	64	56	49	42	35
740	750	143	130	117	105	92	79	67	57	50	43	37
750	760	145	133	120	108	95	82	70	59	52	45	38
760	770	148	136	123	110	98	85	72	60	53	46	40
770	780	151	138	126	113	100	88	75	63	55	48	41
780	790	154	141	129	116	103	91	78	65	56	49	43
790	800	157	144	131	119	106	93	81	68	58	51	44
800	810	159	147	134	122	109	96	84	71	59	52	46
810	820	162	150	137	124	112	99	86	74	61	54	47
820	830	165	152	140	127	114	102	89	77	64	55	49
830	840	168	155	143	130	117	105	92	79	67	57	50
840	850	171	158	145	133	120	107	95	82	69	58	52
850	860	173	161	148	136	123	110	98	85	72	60	53
860	870	176	164	151	138	126	113	100	88	75	62	55
870	880	179	166	154	141	128	116	103	91	78	65	56
880	890	182	169	157	144	131	119	106	93	81	68	58

Name ______________________ Date ____________

Test

Answers

1. Write as a decimal number:
   Two hundred fifty-three and twenty-seven ten thousandths — 1. 253.0027

2. Write 29.71 in words: 2. Twenty-nine and seventy-one hundredths

3. Round 3.075 to the nearest tenth: — 3. 3.1

4. Add: 4.5 + 9.83 + 157.0386 + .195 = — 4. 171.5636

5. Subtract: 14.217 − 3.68 = — 5. 10.537

6. The expenses submitted by a sales representative for a recent trip were: food, $27.63; hotel, $172.25; telephone, $7.71. Calculate the total expense. — 6. $207.59

7. Multiply: 8.532 × 15.731 = — 7. 134.216892

8. Divide: 295.115 ÷ 4.91 =
   (Round to the nearest thousandth) — 8. 60.105

(Test continues on next page)

Test *(Concluded)*

Answers

9. Multiply: a. 4.23 × 10 = 9a. 42.3

b. 4.23 × 100 = 9b. 423

c. 4.23 × 1,000 = 9c. 4,230

10. Divide: a. 19.37 ÷ 10 = 10a. 1.937

b. 19.37 ÷ 100 = 10b. .1937

c. 19.37 ÷ 1,000 = 10c. .01937

11. The state reimburses Bob .33 cents per mile. Bob submits his travel log for a total of 570.6 miles. What does the state pay Bob (round to the nearest cent)? 11. $188.30

12. Convert $16\frac{7}{8}$ to a decimal: 12. 16.875

13. Convert 4.16 to a fraction, reducing to lowest terms: 13. $4\frac{4}{25}$

CHAPTER

4 Percents

UNITS

UNIT

44 Converting Decimals and Fractions to Percents

Percents are widely used in business and in other aspects of life. Working effectively with them requires the understanding that when we speak of percent, we are talking about parts of 100 or division by 100. The symbol used to denote percent is %. Consequently, 25 percent can be written as 25% and is thought of as 25 parts out of 100 parts or $\frac{25}{100}$. Because 25% can be written as $\frac{25}{100}$, it can also be written as a reduced fraction or as a decimal without changing its value.

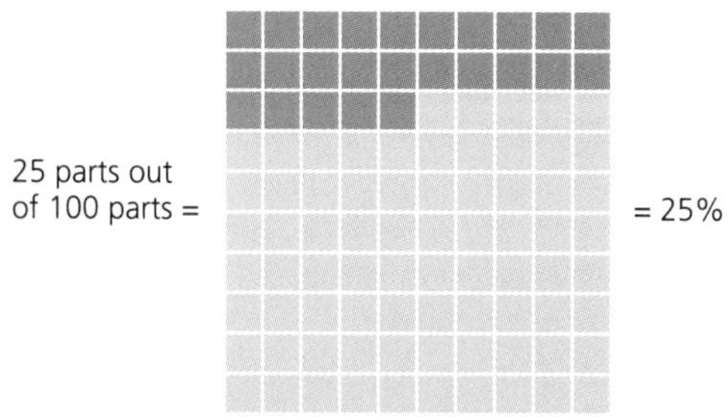

Fractions, decimals, and percents are strongly related, so it is important to understand these relationships and how to convert fractions, decimals, and percents to each of the other forms.

To convert decimals to percents:

1. Rewrite the number and move the decimal point *two places* (remember two places means hundredths) *to the right*. In essence, what we are doing is multiplying by 100.
2. Attach the percent symbol, %, to the number.

To convert fractions to percents:

1. Change any mixed numbers to improper fractions.
2. Multiply the numerator of the fraction by 100.
3. Divide the numerator by the denominator.
4. Round the quotient to the nearest tenth (or whatever place is desired) if the numbers do not divide evenly. The remainder can also be expressed as a fraction where the numerator is the remainder and the denominator is the divisor.
5. Attach the percent symbol to the number.

EXAMPLE 1 **Change .22, .002, and 12.2 to percents.**

Answers:

.22	⇒	move the decimal point two places to the right	=	22%
.002	⇒	move the decimal point two places to the right	=	.2%
12.2	⇒	move the decimal point two places to the right	=	1,220%

PRACTICE 1 **Change .41, .043, and 11.4 to percents.**

Answers: .41 = 41% .043 = 4.3% 11.4 = 1,140%

EXAMPLE 2 **Change $\frac{3}{5}$, $2\frac{1}{8}$, and $\frac{2}{3}$ to percents (round to the nearest tenth percent, as needed).**

$\frac{3}{5} =$ $3 \times 100 = 300$

Answer: $300 \div 5 = 60\%$

$2\frac{1}{8} =$ $2\frac{1}{8} = \frac{17}{8}$

$17 \times 100 = 1{,}700$

Answer: $1{,}700 \div 8 = 212.5\%$

$\frac{2}{3} =$ $2 \times 100 = 200$

$200 \div 3 = 66.66$

Answer: 66.66 rounded to the nearest tenth = 66.7%

PRACTICE 2 **Change $\frac{1}{4}$, $1\frac{1}{4}$, and $\frac{5}{6}$ to percents (round to the nearest tenth percent, as needed).**

Answers: $\frac{1}{4} =$ 25% $1\frac{1}{4} =$ 125% $\frac{5}{6} =$ 83.3%

UNIT **44** Name ______________________ Date ____________

Converting Decimals and Fractions to Percents

Change the following decimals to percents:

1. .04 = 4%
2. .13 = 13%
3. 1.7 = 170%
4. 2.89 = 289%
5. .07 = 7%
6. .302 = 30.2%
7. .0016 = .16%
8. .7 = 70%
9. .2 = 20%
10. .1018 = 10.18%
11. .625 = 62.5%
12. 1.09 = 109%
13. 2.06 = 206%
14. 5 = 500%
15. 3 = 300%
16. 22 = 2,200%

Change the following fractions to percents.

17. $\frac{1}{2}$ = 50%
18. $\frac{1}{5}$ = 20%
19. $\frac{5}{8}$ = 62.5%
20. $3\frac{3}{4}$ = 375%
21. $\frac{3}{5}$ = 60%
22. $\frac{1}{10}$ = 10%
23. $\frac{4}{100}$ = 4%
24. $1\frac{3}{4}$ = 175%

Change the following fractions to percents (round to the nearest tenth percent as needed).

25. $\frac{1}{6}$ = 16.7%
26. $\frac{1}{12}$ = 8.3%
27. $\frac{4}{7}$ = 57.1%
28. $3\frac{1}{7}$ = 314.3%
29. $1\frac{1}{3}$ = 133.3%
30. $\frac{3}{11}$ = 27.3%
31. $1\frac{2}{3}$ = 166.7%
32. $\frac{5}{6}$ = 83.3%

Change the following fractions to percents (express the remainder as a fraction):

33. $\frac{2}{3}$ = $66\frac{2}{3}$%
34. $\frac{7}{9}$ = $77\frac{7}{9}$%
35. $\frac{8}{15}$ = $53\frac{1}{3}$%

Business Applications:

36. In a marketing study conducted by General foods, it was learned that $\frac{1}{5}$ of the college students surveyed do not eat breakfast cereals. Express this fraction as a percent.

20%

37. According to Coca-Cola corporate executives, Coca-Cola currently controls .3 of the Japanese market for carbonated soft drinks. Express this quantity as a percent.

30%

Converting Percents to Fractions and Decimals

The procedures for converting percents to fractions or decimals involve dividing the percent by 100 (or multiplying it by $\frac{1}{100}$). We can acccomplish this operation easily by following these procedures.

To convert percents to decimals:

1. If part of the percent consists of a fraction, express the fraction part as a decimal number.
2. Drop the percent symbol, %.
3. Move the decimal point *two places to the left.*

To convert percents to fractions:

1. If part of the percent consists of a decimal, express the decimal part as a fraction.
2. Change any mixed numbers to improper fractions.
3. Multiply the number by $\frac{1}{100}$.
4. Reduce the fraction to lowest terms.

EXAMPLE 1 **Convert 35%, 3.75% $\frac{1}{4}$%, and $3\frac{1}{4}$% to decimals.**

35% ⇒ Move the decimal point two places to the left and remove the percent symbol (%).

Answer: = .35

3.75% ⇒ Move the decimal point two places to the left and remove the percent symbol (%).

Answer: = .0375

$\frac{1}{4}\%$ ⇒ $\frac{1}{4}\% = .25\%$

.25% ⇒ Move the decimal point two places to the left and remove the percent symbol (%).

Answer: = .0025

$3\frac{1}{4}\%$ ⇒ $3\frac{1}{4}\% = 3.25\%$

3.25% ⇒ Move the decimal point two places to the left and remove the percent symbol (%).

Answer: = .0325

PRACTICE 1 **Change 72%, 6.15%, $\frac{1}{5}$%, and $2\frac{1}{5}$% to decimals.**

Answers: 72% = .72 6.15% = .0615 $\frac{1}{5}\%$ = .002 $2\frac{1}{5}\%$ = .022

EXAMPLE 2 **Change 22%, 1.4%, and $1\frac{3}{5}$% to fractions.**

Answer: $22\% = \quad 22 \times \frac{1}{100} = \frac{22}{100} = \frac{11}{50}$

Answer: $1.4\% = \quad 1\frac{4}{10}\% = \frac{14}{10} \times \frac{1}{100} = \frac{14}{1{,}000} = \frac{7}{500}$

Answer: $1\frac{3}{5}\% = \quad 1\frac{3}{5} \times \frac{1}{100} = \frac{8}{5} \times \frac{1}{100} = \frac{8}{500} = \frac{2}{125}$

PRACTICE 2 **Change 14%, 3.7%, and $2\frac{1}{4}$% to fractions.**

Answers: $14\% = \underline{\frac{7}{50}}$ $\quad 3.7\% = \underline{\frac{37}{1{,}000}}$ $\quad 2\frac{1}{4}\% = \underline{\frac{9}{400}}$

UNIT **45** Name ______________________ Date ____________

Converting Percents to Fractions and Decimals

Convert the following percents to decimals:

1. 11% = .11
2. 6% = .06
3. 27% = .27
4. 114% = 1.14
5. .16% = .0016
6. 1.5% = .015
7. $6\frac{1}{4}$% = .0625
8. $2\frac{1}{8}$% = .02125
9. $\frac{2}{5}$% = .004
10. $\frac{7}{8}$% = .00875
11. $\frac{5}{8}$% = .00625
12. $2\frac{3}{4}$% = .0275
13. $1\frac{1}{2}$% = .015
14. $\frac{5}{16}$% = .003125
15. $\frac{7}{20}$% = .0035
16. 1.08% = .0108
17. $1\frac{3}{16}$% = .011875
18. .0025% = .000025
19. 400% = 4
20. 16.6% = .166

UNIT **45** Name ______________________ Date ______________

Converting Percents to Fractions and Decimals

Convert the following percents to fractions:

21. 25% = $\frac{1}{4}$
22. 39% = $\frac{39}{100}$
23. 6% = $\frac{3}{50}$
24. 142% = $1\frac{21}{50}$
25. 10% = $\frac{1}{10}$
26. 75% = $\frac{3}{4}$
27. 250% = $2\frac{1}{2}$
28. .12% = $\frac{3}{2{,}500}$
29. .7% = $\frac{7}{1{,}000}$
30. 68% = $\frac{17}{25}$
31. $12\frac{1}{2}$% = $\frac{1}{8}$
32. 36% = $\frac{9}{25}$
33. 87.25% = $\frac{349}{400}$
34. 44.5% $\frac{89}{200}$
35. $16\frac{2}{3}$% = $\frac{1}{6}$
36. $4\frac{7}{8}$% = $\frac{39}{800}$
37. $\frac{5}{9}$% = $\frac{1}{180}$
38. $33\frac{1}{3}$% = $\frac{1}{3}$

Business Applications:

39. Massachusetts's sales tax is 5%. Express this value as a decimal.

    .05

40. The U.S. Bureau of Labor Statistics reveals that in the state of Virginia, 6.8% of the workforce is employed by a company doing work strictly for the Department of Defense. Express this percent as a reduced fraction.

    $\frac{17}{250}$

UNIT

46 Percent: Definitions and Formulas

There are three elements to any percent problem, namely, the **base,** the **portion,** and the **rate.** Understanding what these terms mean and being able to identify each is a necessary part of solving any percent problem.

Base— The base is the beginning entire quantity or value with which some other quantity is compared. It is usually the first number following the word *of* in the problem.

Rate— The rate is a percent that indicates what part of the base is being calculated. It is the number followed by the % symbol.

Portion— The portion is the remaining number and is the amount being compared with the base. It can be larger than the base itself if the rate is greater than 100%.

When solving percent problems, two out of three elements are given, and the third must be found. Doing this calculation involves the use of one of these three formulas:

$$\text{Base} = \frac{\text{Portion}}{\text{Rate}}$$

$$\text{Rate} = \frac{\text{Portion}}{\text{Base}}$$

$$\text{Portion} = \text{Rate} \times \text{Base}$$

Notice that solving percent problems involves either multiplication or division.

1. When the portion is unknown ⇒ Multiply the other two quantities.
2. When the portion is known ⇒ Divide the portion by the other known quantity.

The diagram below illustrates these formulas and facts. The horizontal line in the figure is the dividing line, and the vertical line is the multiplying line. To use the figure, place your finger on the element being solved for and perform the operation linking the remaining elements.

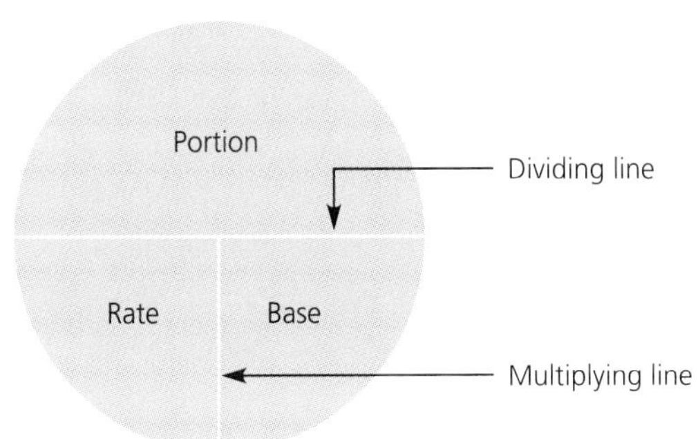

Summarizing all of this information leads us to the following procedure.

1. Identify the base, rate, or portion from the values given in the problem.
2. Select the formula that identifies the unknown element you wish to find.
3. Change the percent to a decimal if necessary.
4. Solve the problem by multiplying or dividing.
5. When solving for the rate, be sure to change the decimal number to a percent.

Remember, some percent problems involve percents greater than 100% or portions greater than the base.

EXAMPLE 1 **Identify the base, rate, and portion in the following problems. If one of the elements is unknown, say so.**

15 is 30% of 50.

Answer: Base = 50 (follows *of*) Rate = 30% Portion = 15

60% of 25 is 15.

Answer: Base = 25 (first number after *of*) Rate = 60% Portion = 15

32 is what percent of 200?

Answer: Base = 200 Rate = unknown Portion = 32

60 is 75% of what number?

Answer: Base = unknown Rate = 75% Portion = 60

PRACTICE 1 **Identify the base, rate, and portion in the following problems. If one of the elements is unknown, say so.**

22% of 200 is 44.

Answer: Base = 200 Rate = 22% Portion = 44

8 is 25% of 32.

Answer: Base = 32 Rate = 25% Portion = 8

What percent of 36 is 12?

Answer: Base = 36 Rate = unknown Portion = 12

13% of 200 is what number?

Answer: Base = 200 Rate = 13% Portion = unknown

EXAMPLE 2 **Solve the following problems.**

What is 75% of 160?

Base = 160 Rate = 75% = .75 Portion = unknown

Use: Portion = Rate × Base
Portion = .75 × 160
Answer: Portion = 120

22 is what percent of 50?

Base = 50 Rate = unknown Portion = 22

Use: $\text{Rate} = \dfrac{\text{Portion}}{\text{Base}}$

$\text{Rate} = \dfrac{22}{50}$

Rate = .44

Answer: Rate = 44%

20% of what number is 13?

Base = unknown Rate = 20% = .20 Portion = 13

Use: $\text{Base} = \dfrac{\text{Portion}}{\text{Rate}}$

$\text{Base} = \dfrac{13}{.20}$

Answer: Base = 65

PRACTICE 2 **Solve the following problems.**

What is 25% of 40? **Answer:** 10

42 is what percent of 1,050? **Answer:** 4%

60% of what number is 252? **Answer:** 420

EXAMPLE 3 **Solve the following problems:**

120% of 90 is what number?

Base = 90 Rate = 120% Portion = unknown

Use: Portion = Rate × Base

Portion = 1.2 × 90

Answer: Portion = 108

72 is 150% of what number?

Base = unknown Rate = 150% Portion = 72

Use: $\text{Base} = \dfrac{\text{Portion}}{\text{Rate}}$

$\text{Base} = \dfrac{72}{1.5}$

Answer: Base = 48

50 is what percent of 20?

Base = 20 Rate = unknown Portion = 50

Use: $\text{Rate} = \frac{\text{Portion}}{\text{Base}}$

$\text{Rate} = \frac{50}{20}$

$\text{Rate} = 2.5$

Answer: $\text{Rate} = 250\%$

PRACTICE 3 **Solve the following problems:**

110% of 70 is what number? **Answer:** 77

48 is 120% of what number? **Answer:** 40

72 is what percent of 36? **Answer:** 200%

UNIT Name ____________ Date ____________

46 *Percent: Definitions and Formulas*

Identify the base, rate, and portion in the following problems (let *b*, *r*, and *p* represent these quantities):

1. 22% of 100 is 22: $b =$ 100 $r =$ 22% $p =$ 22
2. 5 is 20% of 25: $b =$ 25 $r =$ 20% $p =$ 5
3. 45% of 90 is 40.5: $b =$ 90 $r =$ 45% $p =$ 40.5
4. 12% of 80 is 9.6: $b =$ 80 $r =$ 12% $p =$ 9.6
5. 34.2 is 45% of 76: $b =$ 76 $r =$ 45% $p =$ 34.2
6. 90 is 75% of 120: $b =$ 120 $r =$ 75% $p =$ 90
7. 32 is 50% of 64: $b =$ 64 $r =$ 50% $p =$ 32
8. 16.2% of 50 is 8.1: $b =$ 50 $r =$ 16.2% $p =$ 8.1
9. $24\frac{1}{2}\%$ of 200 is 49: $b =$ 200 $r =$ $24\frac{1}{2}\%$ $p =$ 49
10. 66.25 is $13\frac{1}{4}\%$ of 500: $b =$ 500 $r =$ $13\frac{1}{4}\%$ $p =$ 66.25

UNIT **46** Name ______________________ Date ______________

Percent: Definitions and Formulas

Solve the following problems:

11. 3% of 200 is what number? 6
12. What is $5\frac{1}{5}$% of 400? 20.8
13. What is 40% of 840? 336
14. What percent of 12 is 3? 25%
15. 37.5 is what percent of 250? 15%
16. What percent of 14.7 is 9.8? (Round the answer to the nearest tenth percent.) 66.7%
17. 6% of what number is 144? 2,400
18. 3 is $12\frac{1}{2}$% of what number? 24
19. 60% of what number is 24? 40
20. What is 8% of 32? 2.56
21. 1.35 is 3% of what number? 45
22. 150% of 30 is what number? 45
23. 600 is 200% of what number? 300
24. 200 is $62\frac{1}{2}$% of what number? 320
25. 16 is what percent of 64? 25%
26. 8 is what percent of 80? 10%

27. 45% of what number is 486? 1,080

28. What is 70% of 15? 10.5

29. 200 is what percent of 600? (Round the answer to the nearest tenth percent.) 33.3%

30. 1.54 is what percent of 38.5 4%

31. What is $87\frac{1}{2}$% of 1,600? 1,400

32. 112.5 is 45% of what number? 250

33. 46 is what percent of 40? 115%

Business Applications:

34. Delta Airlines said its January traffic rose 8.4% from its January traffic a year earlier of 558,000 passengers. Multiply these numbers to find the portion of passengers that represents this increase.

    46,872 passengers

35. A sweater manufacturer knows that some of the sweaters it stitches will have to be sold as seconds. Over 3 days, it manufactures and stitches 500 sweaters and finds that 52 will be sold as seconds. Divide to express as a percent how many sweaters must be sold as seconds.

    10.4%

How to Dissect and Solve Word Problems: Applications of Percent

Many applications of percent are ones that occur daily, where the base, the rate, or the portion is the unknown quantity. Solving these problems can be successfully done using the formulas presented in Unit 46, along with the aid for dissecting and solving a word problem given in each of the examples that follow.

EXAMPLE 1 **Michael correctly responded to 80% of the 45 questions on a science test. How many questions did he answer correctly?**

Gather the facts.	What am I solving for?	What must I need to know or calculate before solving the problem?	Key points to remember.
Correct responses = 80% Total questions = 45	Number of questions answered correctly	% of correct responses × Total questions = Number of correctly answered questions.	Portion = Rate × Base Change percent to a decimal.

Answer: Portion = Rate × Base = 80% × 45 = .80 × 45 = 36 questions

PRACTICE 1 **O. D. sold 60% of the 50 pieces of furniture he displayed at auction. How many pieces of furniture did he sell?**

Answer: 30 pieces

EXAMPLE 2 **Charlene devotes 15 hours of a 40-hour workweek to administrative responsibilities. What percent of her 40-hour workweek is spent on administrative tasks?**

Gather the facts.	What am I solving for?	What must I need to know or calculate before solving the problem?	Key points to remember.
Total hours worked = 40 Hours spent on administration = 15	Percent of time spent on administrative tasks	Administrative hours ÷ Total hours worked = Percent of time spent on administration	$\text{Rate} = \frac{\text{Portion}}{\text{Base}}$ Change decimal to a percent.

Answer: $\text{Rate} = \frac{\text{Portion}}{\text{Base}} = \frac{15}{40} = .375 = 37.5\%$

PRACTICE 2 **Tony spent 20 hours at his work last week giving cost estimates to his customers. He worked a total of 60 hours. What percent of the time was spent giving estimates? Round the answer to the nearest whole percent.**

Answer: 33%

EXAMPLE 3 **Of the people attending a school fundraiser, 25% won door prizes. If 35 people won prizes, how many people attended the fundraiser?**

Gather the facts.	What am I solving for?	What must I need to know or calculate before solving the problem?	Key points to remember.
% of people who won prizes = 25% # of people winning prizes = 35	The number of people attending the fundraiser	# of people winning prizes ÷ % of people winning prizes = # of people in attendance	Base = $\frac{\text{Portion}}{\text{Rate}}$ Change percent to a decimal.

Answer: $\text{Base} = \frac{\text{Portion}}{\text{Rate}} = \frac{35}{25\%} = \frac{35}{.25} = 140 \text{ people}$

PRACTICE 3 **At a class reunion, 60% of a class attended. How many students are in the class if 270 attended the reunion?**

Answer: 450 students

UNIT 47 Name ______________________ Date ______________

How to Dissect and Solve Word Problems: Applications of Percent

Solve the following problems:

1. Jamil's weight is 72 pounds. Aslam's weight is 90% of Jamil's. What is Aslam's weight?

   64.8 pounds

2. Of the 5 hours a particular course meets during a week, 80% of the time was spent lecturing. How many hours were spent lecturing this particular week?

   4 hours

UNIT **47** Name ______________________ Date ______________

How to Dissect and Solve Word Problems: Applications of Percent

3. Raoul and Sonja spent 85% of the $1,500 they brought on a 2-week pleasure cruise. How much money did they spend?

   $1,275

4. Of the original class of 64 students, 12 dropped out. What percent of the original class dropped out?

   18.75%

5. Of a 125-bed hospital, 80 beds were occupied. What percent of the hospital beds were occupied?

   64%

6. Of the 150 invited guests to a wedding, 30 chose not to attend. What percent of the invited guests chose not to attend?

   20%

7. George paid $13.50 in sales tax on a purchase of $270. What percent of the purchase does the sales tax represent?

   5%

8. Don improved his personal best marathon time of 187 minutes by 8%. By how many minutes did his time improve? (Round to the nearest whole number.)

   15 minutes

9. A $2,000 deposit on a used truck represents a 20% deposit. What is the selling price of the truck?

   $10,000

10. A quality control inspector finds 75, or 2.5%, of the manufactured items being inspected to be defective. How many items were manufactured?

    3,000 items

11. Forty-five members of a Boy Scout troop attended a recent park meeting. If this number represented 60% of the total membership, what is the total membership?

    75 Boy Scouts

12. Of the signatures obtained by a candidate for public office, 8% were invalidated. If 52 of the total signatures were invalidated, what was the total number of signatures collected?

    650 signatures

13. Out of 18 times at bat, Colleen hit 4 singles. What percent of the time at bat did she hit singles? (Round to the nearest tenth.)

22.2%

14. Out of 60 true-false questions on Bert's test, 24 were answered incorrectly. What percent of the true-false questions did he answer incorrectly?

40%

15. Of the 25 windows in Sarom's house, 40% need to be replaced. How many windows need replacing?

10 windows

16. Early retirement incentives were taken advantage of by 20% of the workers in a local municipality. If 62 employees took advantage of this early retirement plan, how many total employees are there?

310 employees

17. Of Catherine's weekly salary, 27% goes to taxes, health insurance, union dues, and miscellaneous other deductions. If her total deductions add up to $283.50, what is her weekly salary?

$1,050

18. A local junior college has collected donations of $156,000 from its alumni. If this amount represents 25% of its target amount, what is its target amount?

$624,000

19. A survey of 500 members of a health maintenance organization revealed that 25% were dissatisfied with its services. How many members were dissatisfied?

125 members

20. Out of 36 students in our Algebra class, 25% received *A*'s and 50% received *B*'s or *C*'s. The rest of the students received either *D*'s, *F*'s, or *W*'s. How many students received *D*'s, *F*'s, or *W*'s?

9 students

How to Dissect and Solve Word Problems: Percent of Increase and Decrease

Many times when solving for the base, rate, or portion in a percent problem, an increase or decrease of a quantity has occurred. Solving these kinds of problems means we will be trying to find (1) the amount of increase or decrease, (2) the rate of increase or decrease, (3) the base quantity before the increase, or (4) the quantity after the increase.

EXAMPLE 1 **Total retail sales for Cheap Fashions this year decreased 8% from its total sales of $875,550 last year. What is this year's total retail sales figure?**

Gather the facts.	What am I solving for?	What must I need to know or calculate before solving the problem?	Key points to remember.
% decrease = 8% Last year's sales = $875,550	This year's total retail sales figure	Last year's total sales – (% decrease × Last year's total sales) = This year's total retail sales	Portion = Rate × Base

Answer: This year's total retail sales = $875,550 – (.08 × $875,550)
= $875,550 – ($70,044)
= $805,506

Alternative solution: Since sales decreased 8% from last year, this year's sales are 92% of last year's sales.

92% of $875,550 = .92 × $875,550 = $805,506

PRACTICE 1 **The Roberts family's electric bill for January decreased by 15% over the December bill of $70.85. What was January's bill? (Round the answer to the nearest cent.)**

Answer: *$60.22*

EXAMPLE 2 **A local school increased its tuition from $1,700 per year to $1,850 per year. What percent of the original tuition is the increase? (Round to the nearest tenth percent.)**

Gather the facts.	What am I solving for?	What must I need to know or calculate before solving the problem?	Key points to remember.
Last year's tuition = $1,700 This year's tuition = $1,850	The percent of increase	(This year's tuition – Last year's tuition) ÷ Last year's tuition = Percent of increase	Rate = $\frac{\text{Portion}}{\text{Base}}$

Answer: Percent of increase = [$1,850 – $1,700] ÷ $1,700 = [$150 ÷ $1,700]
= .0882 = 8.8%

PRACTICE 2 **The population of Bay Ridge decreased to 23,779 from its population of 25,387 ten years ago. What is the percent of decrease? (Round to the nearest tenth percent.)**

Answer: 6.3%

EXAMPLE 3 **Enrollment in real estate courses this semester in Essex Community College has declined 30% from last semester's enrollment of 140 students. What is the decrease in enrollment?**

Gather the facts.	What am I solving for?	What must I need to know or calculate before solving the problem?	Key points to remember.
Last year's enrollment = 140 students % of decrease = 30%	Decrease in enrollment	% of decrease × Last year's enrollment = Amount of decrease	Portion = Rate × Base

Answer: Decrease in enrollment = .30 × 140 = 42 students

PRACTICE 3 **The number of students applying to a pre-engineering science program at a state college increased 28% over last year's 175 applicants. By what amount did applications increase?**

Answer: 49 applications

EXAMPLE 4 **Rent on an apartment was increased $60 per month, which represents an 8% increase. What was the monthly rent before the increase?**

Gather the facts.	What am I solving for?	What must I need to know or calculate before solving the problem?	Key points to remember.
% of increase = 8% Amount of increase = $60	Monthly rent before the increase	Amount of increase ÷ % of increase = Rent before increase	Base = $\frac{\text{Portion}}{\text{Rate}}$

Answer: Rent before increase = $60 ÷ .08 = $750

PRACTICE 4 **A baseball team reduces the number of games it plays by 9 games, which represents a 6% decrease. How many games were played before the decrease?**

Answer: *150 games*

UNIT **48** Name ____________________ Date ____________

How to Dissect and Solve Word Problems: Percent of Increase and Decrease

Solve the following problems:

1. The price of generic drugs rose by 22%. The old price was $16. What is the new price?

   $19.52

2. The number of people voting in a general municipal election increased by 1,120. If 6,862 people voted in the primary election, what is the percent of increase of voter turnout? (Round the answer to the nearest tenth percent.)

   16.3%

3. In one year the value of a $30,000 investment decreased $1,500. What was the percent decrease?

   5%

4. A condo bought for $63,000 has increased in value to $84,000. Find the percent of increase (Round the answer to the nearest tenth percent.)

   33.3%

5. A real estate office handled sales of 256 houses this past year. They wish to increase sales by 15% next year. Approximately how many more sales must the office handle? (Round the answer to the next whole number.)

   39 houses

6. Tim increased his sales of golf equipment by 25% from $105,000. What were his total sales?

   $131,250

7. A regional chamber of commerce reports that there was a 16% decrease in visitors to area tourist attractions this year from last year's record of 11,500 visitors. How many tourists visited attractions this year?

   9,660 tourists

8. After exercising, a man's normal pulse rate of 64 beats per minute increases by 28 beats per minute. What percent of his normal pulse rate was the increase?

   43.75%

9. The seating capacity of an arena is increased by 2,200 seats, which represents a 20% increase. What was the arena's former seating capacity?

   11,000 seats

10. Gary recently sold his house for $135,000. This sale represented a 15% loss off the original price of the house. What was the original price of the house? (Round the answer to the nearest whole dollar.)

    $158,824

11. Michelle managed to deposit $315 into a savings account last month. She deposited $375 into the same account this month. What was the percent increase in savings? (Round the answer to the nearest tenth percent.)

    19%

12. Carol received a salary increase of 6%. Her previous salary was $34,933. What is the amount of increase she received?

    $2,095.98

13. Gordon's savings account balance is $720. His grandfather promised him he would increase the balance by 7%. What will be the new balance?

    $770.40

14. A family's monthly food expense has decreased by $80 since their son went away to college, which is a 32% decrease. How much money per month was spent on food before the decrease?

    $250

15. Current enrollment in adult education courses has increased by 25% over its previous enrollment. If the current enrollment totals 625 students, what was the previous enrollment?

    500 students

16. A certified public accountant has increased the number of clients she sees by 40%. If she has taken on 52 new clients, how many clients did she previously have?

    130 clients

17. Winds atop Mount Washington were reported to have increased in velocity by 72% during 1 hour. At the start of the hour they were measured at 18 miles per hour. What did they measure after the 72% increase?

    30.96 mph

18. John has decided to increase the size of his garden from 360 square feet to 625 square feet. What is the percent of increase? (Round the answer to the nearest tenth percent.)

    73.6%

19. A car manufacturer has announced an increase of 2.5% for its new model cars. The current price of one of these cars is $16,899. What will its price be after the increase? (Round the answer to the nearest cent.)

    $17,321.48

20. Retail sales for this month were reported down by 15% from last month. Last month's sales totaled $177,685. What are the sales for this month?

    $151,032.25

UNIT 49 How to Dissect and Solve Word Problems: Commission Sales

Many businesses pay their sales personnel a **commission,** a percent of the total dollar value of the sale. The definition and formulas used for solving percent problems apply to commission sales problems also. The following formulas are similar to the formulas presented in Unit 46 and can be used to solve the percent problems involving commission in this unit.

Amount of commission earned = Rate of commission × Sales

Rate of commission = Amount of commission earned ÷ Sales amount

Sales amount = Amount of commission earned ÷ Rate of commission

EXAMPLE 1 **Andrew sold $16,500 worth of computer hardware to a private school. His commission rate is 11%. What is the amount of commission he earned?**

Gather the facts.	What am I solving for?	What must I need to know or calculate before solving the problem?	Key points to remember.
Commission rate = 11% Amount of sales = $16,500	Amount of commission earned	Commission rate × Sales amount = Amount of commission	Move decimal two places to the left.

Answer: Amount of commission = 11% × $16,500 = .11 × $16,500 = $1,815

PRACTICE 1 **Don is paid a commission of 5% on each house he sells. He recently sold a house for $145,900. What is his commission?**

Answer: $7,295

EXAMPLE 2 **Sandy is paid a commission of 7.5% on sales in excess of $150,000 plus a base salary of $2,400 a month. Her total sales amounted to $185,000 last month. What was her salary?**

Gather the facts.	What am I solving for?	What must I need to know or calculate before solving the problem?	Key points to remember.
Base salary = $2,400 Total sales = $185,000 Commission rate = 7.5%	Monthly salary	(Sales in excess of $150,000 × Commission rate) + Base salary = Month's salary	Determine sales in excess of $150,000.

Find sales in excess of $150,000: $185,000 – $150,000 = $35,000

Commission = 7.5% × $35,000 = .075 × $35,000 = $2,625

Answer: Month's salary = $2,625 + $2,400 = $5,025

PRACTICE 2 **Carolyn earns a 5% commission on all sales in excess of $7,500 plus a base salary of $75 per week. Her total sales this week amount to $10,550. Calculate her week's salary.**

Answer: *$227.50*

EXAMPLE 3 **A company offers to pay a salesperson a 5% commission. How much merchandise must the salesperson sell to receive a commission of $1,200?**

Gather the facts.	What am I solving for?	What must I need to know or calculate before solving the problem?	Key points to remember.
Commission rate = 5% Amount of commission = $1,200	Sales amount	Amount of commission ÷ Rate of commission = Sales amount.	Change percent to a decimal.

Answer: $\text{Sales} = \frac{\$1{,}200}{5\%} = \frac{\$1{,}200}{.05} = \$24{,}000$

PRACTICE 3 **To receive a commission of $750 a salesperson must sell $12,500 worth of goods. What is the rate of commission being paid?**

Answer: *6%*

UNIT **49** Name ____________ Date ____________

How to Dissect and Solve Word Problems: Commission Sales

Solve the following problems:

1. A real estate broker sold a house for $219,500. The commission was 6% of the sales price. What commission did the broker receive?

   $13,170

2. Jeannie receives a base salary of $5,500 a year plus 3.5% of her total sales. Her total sales last year were $64,000. What was her yearly salary?

   $7,740

3. Miguel receives an 8% commission on all sales. During the month of January, his total sales were $16,893. What was his commission?

   $1,351.44

4. The selling price of a house was $156,500, and the commission due the realtor amounted to $10,955 of this price. What was the rate of commission charged?

   7%

5. Marco Realty received a commission of $27,502.50 on a piece of property that sold for $289,500. What was the rate of commission?

   9.5%

6. Arthur Conner sells automobiles and receives a salary of $250 per week plus a 3.5% commission on all sales. His total sales for one week were $17,300. What was his weekly salary?

   $855.50

7. René works for a company that pays her a base salary of $200 per week plus a commission of 4.2% on weekly sales in excess of $4,800. Her total sales over four weeks amounted to $35,200. What did her salary amount to for these four weeks?

   $1,472.00

8. Calculate the weekly salary for a salesperson who earns a base salary of $600 plus a 3% commission on sales totaling $7,500.

   $825

9. What is a realtor's sales commission if a condominium sells for $68,950 and the rate of commission is 7.5% of the sales price?

   $5,171.25

10. A salesperson's commission is 5%. How much merchandise must be sold to earn $1,625?

    $32,500

11. An agency pays its sales personnel a weekly salary of $170 plus 4% commission on all sales. What must the sales amount be if a weekly salary of $650 is desired?

$12,000

12. Peter Berwick received a $200 commission on the sale of $5,000 worth of goods. What rate of commission did he receive?

4%

13. Cyr Associates pays its sales force graduated commissions. Each salesperson is paid a 2% commission on sales up to $10,000; 3% on the next $5,000; and 4.5% on all sales over $15,000. Calculate the salary on $25,000 worth of sales.

$800

14. Parker earns 10% commission an all sales plus 5% on sales over $6,000. His total sales were $6,290. What was his total commission?

$643.50

15. J. W. Person is paid 6% commission on the first $15,000 of monthly sales and 8% on all sales in excess of $15,000. His sales for the month totaled $32,700. What was his total commission for the month?

$2,316

16. Vin Crivic is paid a weekly salary of $300, along with a 5% commission on all sales over $12,000 a week. What must be his total weekly sales if his total weekly income is to be $1,400?

$34,000

17. A salesperson's commission is $3\frac{1}{2}$% on all sales. What must be the amount of the sales to have earned $1,057.70?

$30,220

18. Sally Chase is employed on a salary plus commissions basis. Her total earnings for last year amounted to $63,700 of which $16,500 represented her guaranteed annual salary. Her total sales last year amounted to $590,000. What percent commission did she receive?

8%

19. Paul Lospellato received a $7,241 commission on a house that sold for $144,820. What was his rate of commission?

5%

20. A 6% commission of $10,650 was earned by Thorndike Realty on the sale of a 2-family house. What was the house's selling price?

$177,500

UNIT

50 Trade Discounts

A **trade discount** is a reduction of the original selling price of an item that retailers receive from manufacturers. The original selling price is referred to as the **list price** and the price the retailer pays to the manufacturer after having deducted the trade discount is called the **net price.** When manufacturers give trade discounts they do so by offering a single trade discount rate or a series of two or more trade discount rates called a **chain discount.** When offering a chain discount, the manufacturers list them as a group. For example, a series of trade discounts of 20%, 15%, and 10% would be stated as 20/15/10. Trade discounts may be computed by multiplying the *trade discount rate and the list price to arrive at the trade discount amount. Subtracting the trade discount amount from the list price gives the net price.*

When a chain discount is given, the long way to calculate the net price would be to multiply the trade discount rate and the new balance after subtracting the previous trade discount amount. A more effective way of finding the net price would be to:

1. Subtract each trade discount rate from 100% to obtain the complement of the trade discount rate. Convert each to a decimal.
2. Multiply the decimals to find the net price rate. *Do not round off.*
3. Multiply the list price and the net price rate to obtain the net price.

EXAMPLE 1 **Find the amount of discount and the net price.**

List Price	Trade Discount Rate	Trade Discount	Net Price
$1,500	7.5%		

Answers: Trade Discount = .075 × $1,500 = $112.50

Net Price = $1,500 − $112.50 = $1,387.50

PRACTICE 1 **Find the amount of discount and the net price.**

List Price	Trade Discount Rate	Trade Discount	Net Price
$2,800	12%	$336	$2,464

EXAMPLE 2 **Complete the following table:**

List Price	Chain Discount	Net Price Rate	Net Price	Trade Discount
$3,500	10/5/2			

Net Price Rate = 100% − 10% = 90%; 100% − 5% = 95%; 100% − 2% = 98%

(90%, 95%, 98%: Complement of Trade Discount Rates)

Answers: Net price rate = .9 × .95 × .98 = .8379

Net price = .8379 × $3,500 = $2,932.65

Trade discount = $3,500 − $2,932.65 = $567.35

PRACTICE 2 **Complete the following table:**

List Price	Chain Discount	Net Price Rate	Net Price	Trade Discount
$15,000	15/10/5	.72675	$10,901.25	$4,098.75

EXAMPLE 3 **A furniture piece's list price is $1,200 with a trade discount of 11%. What is the net price of the furniture?**

Gather the facts.	What am I solving for?	What must I need to know or calculate before solving the problem?	Key points to remember.
List price = $1,200 Trade discount = 11%	Net price	Net price rate × List price = Net price	Subtract trade discount from 100% to find net price rate.

Answer: Net price rate = 100% − 11% = 89% = .89

Net price = $1,200 × .89 = $1,068

PRACTICE 3 **Efficient Energy Systems buys coal stoves from a wholesaler. Each stove has a list price of $700 with a trade discount of 25%. What is the net price of the stove?**

Answer: $525

UNIT **50** Name ______________________ Date ______________

Trade Discounts

Complete the following tables:

1. Find the trade discount and net price.

List Price	Trade Discount Price	Trade Discount	Net Price
$1,250	40%	$500.00	$750.00
$1,500	30%	$450.00	$1,050.00
$6,005	45%	$2,702.25	$3,302.75
$1,850	15%	$277.50	$1,572.50
$5,775	35%	$2,021.25	$3,753.75
$680	22%	$149.60	$530.40
$2,780	16.5%	$458.70	$2,321.30
$3,750	7.5%	$281.25	$3,468.75
$4,900	$13\frac{1}{4}$%	$649.25	$4,250.75
$7,450	$18\frac{1}{2}$%	$1,378.25	$6,071.75

2. Find the net price rate, net price, and trade discount.

List Price	Chain Discount	Net Price Rate	Net Price	Trade Discount
$4,500	20/10	.72	$3,240.00	$1,260.00
$3,250	25/10	.675	$2,193.75	$1,056.25
$5,775	30/20	.56	$3,234.00	$2,541.00
$1,865	25/10	.675	$1,258.88	$606.12
$18,000	10/5/2	.8379	$15,082.20	$2,917.80
$25,000	20/15/10	.612	$15,300.00	$9,700.00
$12,000	25/20/5	.57	$6,840.00	$5,160.00
$15,000	10/8/3	.80316	$12,047.40	$2,952.60
$105,000	20/15/12	.5984	$62,832.00	$42,168.00
$32,000	20/15/10	.612	$19,584.00	$12,416.00

Solve the following problems:

3. A computer lists for $1,200 with a trade discount of 22%. What is the net price of the computer?

$936.00

4. Curtis Bookseller paid a $4,500 net price for textbooks. The publisher offered a 25% trade discount. What was the publisher's list price?

$6,000

5. Quality Furniture bought a sofa for the net price of $900.00. The sofa had a list price of $1,200. What was the trade discount rate?

25%

6. A roller skates manufacturer offered a 5/2/1 chain discount to its customers. Bob's Sporting Goods ordered roller skates for a total $625 list price. What was the net price of the roller skates? What was the trade discount? (Round the answer to the nearest cent.)

Net Price = $576.06 Trade Discount = $48.94

7. Quality Furniture paid $90 for a lamp after a chain discount of 30/20. What was the list price? (Round the answer to the nearest cent.)

$160.71

8. The Sandstorm Bookstore paid a $6,600 net price for novels it ordered from a publisher. The publisher's list price was $8,250. What was the trade discount rate?

20%

9. Jebco Manufacturing sold Foraly Hardware a set of lawn furniture with a $1,350 list price. Jebco offered a 5/4/2 chain discount. What was the net price of the lawn furniture? (Round the answer to the nearest cent.)

$1,206.58

10. Clean Air Corp. purchases air conditioners at a list price total of $4,800. A series of trade discounts of 25/15/5 are offered. What is the net price?

$2,907.00

11. Donohue Construction plans to purchase construction equipment with a $7,200 list price with a chain discount of 25/10/5. What is the trade discount? What is the net price?

Trade Discount = $2,583.00 Net Price = $4,617.00

12. What is the net price of 8 watches purchased by a jeweler at a list price of $87.50 each, along with a chain discount of 15/10/5? (Round the answer to the nearest cent.)

$508.73

UNIT 51 Cash Discounts

Sellers frequently offer cash discounts to buyers if the invoice price is paid within a specified time. For example, if the terms of the sale are stated as 2/10, n/30, buyers can take a 2% cash discount off the amount of the invoice if the bill is paid within 10 days from the invoice date. If the discount period is missed, the buyers must pay the net amount without a discount between day 11 and day 30 from the date of the invoice.

Similarly, if 2/10, 1/15, n/30 are the terms of the sale, the 2% discount can be taken within 10 days of the date of the invoice. If the 2% discount is missed, a 1% discount can be taken from day 11 to day 15. If this payment is not made on day 15, the full amount is due 30 days from the invoice date.

EXAMPLE 1 **Complete the following table.**

Date of Invoice	Amount of Invoice	Terms	Date Paid	Cash Discount	Net Amount Paid
January 13	$1,080	2/10, n/30	January 19		

Answers: End of 2% discount period:

January 13
+ 10 days
January 23 ends 2% discount period

Payment on January 19:

Cash discount = .02 × $1,080 = $21.60

Net amount paid = $1,080 − $21.60 = $1,058.40

PRACTICE 1 **Complete the following table.**

Date of Invoice	Amount of Invoice	Terms	Date Paid	Cash Discount	Net Amount Paid
September 3	$1,460	2/10, n/30	September 7	$29.20	$1,430.80

EXAMPLE 2 **Complete the following table.**

Date of Invoice	Amount of Invoice	Terms	Date Paid	Cash Discount	Net Amount Paid
March 12	$950	3/10, 2/15, n/30	March 25		

Answers: End of 3% discount period:

March 12
+ 10 days
March 22 ends 3% discount period

End of 2% discount period:

March 22 ends 3% discount period
+ 5 days
March 27 ends 2% discount period

Payment on March 25:

Cash discount = .02 × $950 = $19

Net amount paid = $950 − $19 = $931

PRACTICE 2 **Complete the following table.**

Date of Invoice	Amount of Invoice	Terms	Date Paid	Cash Discount	Net Amount Paid
July 18	$2,000	3/10, 2/15, n/30	July 29	$40.00	$1,960.00

UNIT 51 Name ______________________ Date ____________

Cash Discounts

Assuming that each invoice is paid within the cash discount period, complete the following table and round to the nearest cent as necessary:

Amount of Invoice	Terms	Cash Discount	Net Amount Paid
$1,995	3/10, n/30	$59.85	$1,935.15
$2,025	2/15, n/30	$40.50	$1,984.50
$3,199	4/10, n/30	$127.96	$3,071.04
$4,220	2/10, n/30	$84.40	$4,135.60
$1,650.45	5/10, n/30	$82.52	$1,567.93
$1,735.90	2/15, n/45	$34.72	$1,701.18
$5,746	2/10, n/30	$114.92	$5,631.08
$916	3/10, n/45	$27.48	$888.52
$2,488	4/10, n/30	$99.52	$2,388.48
$649	5/10, n/30	$32.45	$616.55

UNIT 51 Name ______________________ Date ____________

Cash Discounts

Complete the following table:

Date of Invoice	Amount of Invoice	Terms	Date Paid	Cash Discount	Net Amount Paid
April 13	$4,800	3/10, n/30	April 18	$144.00	$4,656.00
February 3	$1,600	5/10, n/30	February 28	0	$1,600.00
June 6	$2,946	2/10, 1/15, n/30	June 18	$29.46	$2,916.54
October 12	$626	3/10, 2/15, n/30	October 25	$12.52	$613.48
March 26	$1,980	1/10, n/30	April 4	$19.80	$1,960.20
October 10	$4,770	2/10, 1/30, n/60	October 28	$47.70	$4,722.30
January 15	$2,015	3/5, 2/20, n/60	February 2	$40.30	$1,974.70
November 3	$1,675	2/10, 1/15, n/30	November 17	$16.75	$1,658.25
August 17	$1,222	3/10, 2/15, n/30	August 31	$24.44	$1,197.56
September 12	$850	2/10, 1/15, n/30	September 19	$17.00	$833.00
July 27	$4,500	2/10, 1/15, n/30	August 6	$90.00	$4,410.00
November 24	$6,750	3/10, 2/15, n/30	December 5	$135.00	$6,615.00
June 13	$7,800	2/10, 1/30, n/60	July 12	$78.00	$7,722.00
August 21	$844	5/50, 2/30, n/60	September 17	$16.88	$827.12
April 6	$5,200	2/10, 1/30, n/60	May 26	0	$5,200.00

Solve the following problems:

1. McGee of New York sold Jolly of Chicago office equipment with a $6,000 list price. Sale terms were 3/10, n/30. Jolly pays the invoice within the discount period. What does Jolly pay McGee?

   $5,820.00

2. Maplewood Supply received a $5,250 invoice dated April 15, 1993. Terms were 4/10, 3/30, n/60. Maplewood pays the invoice on April 27. What does it pay?

   $5,092.50

3. For problem 2, what will Maplewood pay if it pays the invoice on May 21, 1993?

   $5,250.00

4. Hiawatha Supply received an invoice dated April 15, 1993, with a balance of $5,500. Terms were 4/10, 3/30, n/60. Hiawatha pays the invoice on April 29. What amount does Hiawatha pay?

   $5,335.00

5. Majestic Manufacturing sold McCormack Furniture a room set for $8,500. Terms were 3/10, n/30, and the invoice was dated May 30. McCormack paid the invoice on June 9. What was the amount paid by McCormack?

   $8,245.00

UNIT 52 Simple Interest

Interest is the amount of money paid for the use of another person's/organization's money. Terms frequently used when speaking about interest are **principal, rate, time,** and **simple interest. Principal** refers to the amount borrowed or deposited. **Rate** refers to the percentage rate charged or offered, and it is expressed as an annual rate unless stated otherwise. **Time** refers to the period for which the money is borrowed or deposited, and it is usually expressed as the fractional part of a year or as whole years. **Simple interest** refers to the amount of money paid, *based only on the original principal.*

The simple interest formula is:

Interest = Principal × Rate × Time

When using this formula, remember that:

1 Rate should be expressed as a decimal.

2 Time should be expressed as the fractional part of a year or as whole years. For example:

a. $\text{Time} = \dfrac{\text{Exact number of days}}{360^*}$

b. $\text{Time} = \dfrac{\text{Exact number of months}}{12}$

c. Time = Exact number of years

The simple interest formula can also take these forms.

$$\text{Principal} = \frac{\text{Interest}}{\text{Rate} \times \text{Time}}$$

$$\text{Rate} = \frac{\text{Interest}}{\text{Principal} \times \text{Time}}$$

$$\text{Time} = \frac{\text{Interest}}{\text{Principal} \times \text{Rate}}$$

EXAMPLE 1 **Fill in the blanks of the following table.**

	Interest	Principal	Rate	Time
A		$2,000	8%	90 days
B	$450		7.5%	2 years
C	$42.50	$1,700		3 months
D	$900	$7,200	5%	

Answer: A. $\text{Interest} = \text{Principal} \times \text{Rate} \times \text{Time} = \$2{,}000 \times .08 \times \dfrac{90}{360} = \$40$

Answer: B. $\text{Principal} = \dfrac{\text{Interest}}{\text{Rate} \times \text{Time}} = \dfrac{\$450}{.075 \times 2} = \$3{,}000$

* The use of 360 days is done to simplify calculations. It is sometimes referred to as the Banker's rule, but its use in banks today has been curtailed because computer use has made simplified "hand" calculations unnecessary.

Answer: C. $\text{Rate} = \dfrac{\text{Interest}}{\text{Principal} \times \text{Time}} = \dfrac{\$42.50}{\$1{,}700 \times \frac{3}{12}} = \dfrac{\$42.50}{\$425} = 10\%$

Answer: D. $\text{Time} = \dfrac{\text{Interest}}{\text{Principal} \times \text{Rate}} = \dfrac{\$900}{\$7{,}200 \times .05} = 2.5 \text{ years}$

PRACTICE 1 **Fill in the blanks of the following table.**

	Interest	Principal	Rate	Time
A	$18.75	$3,750	6%	30 days
B	$240	$1,600	5%	3 years
C	$32	$800	8%	6 months
D	$187.50	$5,000	2.5%	1.5 years

EXAMPLE 2 **Ernie paid $7.50 interest on a loan of $900 for 60 days. What rate of interest did he pay?**

Gather the facts.	What am I solving for?	What must I need to know or calculate before solving the problem?	Key points to remember.
Interest = $7.50 Principal = $900 Time = 60 days	Rate of interest	$\text{Rate} = \dfrac{\text{Interest}}{\text{Principal} \times \text{Time}}$	Express time as a fractional part of a year $(\frac{n}{360})$.

Answer: $\text{Rate} = \dfrac{\text{Interest}}{\text{Principal} \times \text{Time}} = \dfrac{\$7.50}{\$900 \times \frac{60}{360}} = \dfrac{\$7.50}{\$900 \times \frac{1}{6}} = \dfrac{\$7.50}{\$150} = .05 = 5\%$

PRACTICE 2 **Patrick repaid his $1,300 loan in seven months at an interest rate of 9%. Find the amount of interest paid.**

Answer: $68.25

UNIT 52 Name ______________________ Date ______________

Simple Interest

Fill in the blanks of the following table:

Interest	Principal	Rate	Time
$7.25	$1,450	3%	60 days
$480	$4,000	6%	2 years
$227.50	$3,500	$6\frac{1}{2}$%	1 year
$1,350	$3,750	12%	3 years
$98.80	$760	$3\frac{1}{4}$%	4 years
$20.00	$1,000	6%	120 days
$800	$2,500	8%	4 years
$9.00	$800	$4\frac{1}{2}$%	90 days
$1,350	$3,750	12%	3 years
$26.25	$125	7%	3 years
$41.25	$250	5.5%	3 years
$148.50	$2,140.54	$9\frac{1}{4}$%	9 months
$100.00	$4,000	5%	6 months
$495	$1,500	$5\frac{1}{2}$%	6 years
$6	$240	5%	6 months

Solve:

1. Barry's Ace Construction firm borrowed \$9,500 for four years at a rate of $8\frac{1}{4}$%. Calculate the interest owed on this loan.

\$3,135

2. Tonya's original investment of \$2,420 in a special growth account earned interest totaling \$217.80 after one year. Find the interest rate.

9%

3. The interest for one year on Bill's home improvement loan amounts to \$1,344. The interest rate charged is 16%. Find the amount of the loan.

\$8,400

4. Arcante Plumbing has calculated the interest charge on the \$120,500 it had borrowed to be \$33,740. The term of the loan is 2 years. What is the interest rate charged?

14%

5. The interest due on Randy's car loan amounts to \$217.80 at the end of 1 year, and the interest rate charged is 9%. What is the amount of principal?

\$2,420

6. Mike invested \$1,500 into a 1-year CD for his daughter Ashley. The total interest amounts to \$93.75. What was the rate of interest on the CD?

6.25%

7. How much money must Diane invest for one year at $5\frac{3}{4}$% simple interest to earn \$350 worth of interest? (Round to the nearest dollar.)

\$6,087.00

8. Find the amount of interest that will have accrued on a \$1,200 loan at 11% per year after 5 months.

\$55

9. If Karl takes out a 4-year, \$6,000 car loan at 7%, calculate his total interest payment.

\$1,680

10. Brice took out a loan of \$2,200 at $8\frac{1}{4}$%. What interest will he have to pay at the end of 18 months?

\$272.25

UNIT

53 Compound Interest

Interest calculated on the original principal plus the interest of a prior period is called **compound interest.** When compounding, interest calculated at the end of a period is added to the beginning principal of the period. This sum becomes the new principal on which interest for the next period is figured. This process repeats itself as many times as there are **compounding periods;** thus, compounding results in higher total interest than does simple interest. The total obtained at the end of the last period is called the **compound amount** and the difference between the compound amount and the original principal is the **compound interest.**

Compounding interest is usually done annually, semiannually (every 6 months), quarterly (every 3 months, or daily. It is important, therefore, to find the interest rate per compounding period. This calculation is done by dividing the annual interest rate by the number of compounding periods.

Let's look at an example where $1,000 will earn interest at the rate of 4% compounded annually for 2 years.

Year 1

$1,000 × .04 × 1 = $40	Interest of first period
$1,000 + $40 = $1,040	Beginning Principal for year 2

Year 2

$1,040 × .04 × 1 = $41.60	Interest of second period
$1,040 + $41.60 = $1,081.60	Compound amount
$1,081.60 – $1,000 = $81.60	Compound interest

Although compound interest can be calculated by this "summing" method described above, more often used is a compound interest table. This table gives the compound amounts for a $1 deposit held for a given length of time at a given rate.

To find a compound amount and the compound interest:

1. Determine the interest rate per compounding period by dividing the annual rate by the number of compounding periods.

   Annually—1 period per year

   Semiannually—2 periods per year

   Quarterly—4 periods per year
2. Determine the number of periods over which the compounding occurs by multiplying the number of years and the number of times compounded per year.
3. Find the table's factor for the interest rate per compounding period and the number of periods.
4. Multiply the factor and the principal to obtain the compound amount.
5. Subtract the original principal from the compound amount to find the compound interest.

Table 4.53 on page 170 is a compound interest table. To use this compound interest table:

1. Move down the left-hand column to the desired period number and across to the desired percent.
2. The number at the intersection of the period column and the row of percents is the desired table factor.

For example, if the number of periods was 4 and 8% was the rate, the appropriate table factor would be 1.3605.

Use Table 4.53 to solve the practice problems and exercises in this unit.

Table 4.53
Future Value of $1 at Compound Interest

Period	1%	$1\frac{1}{2}$%	2%	3%	4%	5%	6%	7%	8%	9%	10%
1	1.0100	1.0150	1.0200	1.0300	1.0400	1.0500	1.0600	1.0700	1.0800	1.0900	1.1000
2	1.0201	1.0302	1.0404	1.0609	1.0816	1.1025	1.1236	1.1449	1.1664	1.1881	1.2100
3	1.0303	1.0457	1.0612	1.0927	1.1249	1.1576	1.1910	1.2250	1.2597	1.2950	1.3310
4	1.0406	1.0614	1.0824	1.1255	1.1699	1.2155	1.2625	1.3108	1.3605	1.4116	1.4641
5	1.0510	1.0773	1.1041	1.1593	1.2167	1.2763	1.3382	1.4026	1.4693	1.5386	1.6105
6	1.0615	1.0934	1.1262	1.1941	1.2653	1.3401	1.4185	1.5007	1.5869	1.6771	1.7716
7	1.0721	1.1098	1.1487	1.2299	1.3159	1.4071	1.5036	1.6058	1.7138	1.8280	1.9487
8	1.0829	1.1265	1.1717	1.2668	1.3686	1.4775	1.5938	1.7182	1.8509	1.9926	2.1436
9	1.0937	1.1434	1.1951	1.3048	1.4233	1.5513	1.6895	1.8385	1.9990	2.1719	2.3579
10	1.1046	1.1605	1.2190	1.3439	1.4802	1.6289	1.7908	1.9672	2.1589	2.3674	2.5937
11	1.1157	1.1780	1.2434	1.3842	1.5395	1.7103	1.8983	2.1049	2.3316	2.5804	2.8531
12	1.1268	1.1960	1.2682	1.4258	1.6010	1.7959	2.0122	2.2522	2.5182	2.8127	3.1384
13	1.1381	1.2135	1.2936	1.4685	1.6651	1.8856	2.1329	2.4098	2.7196	3.0658	3.4523
14	1.1495	1.2318	1.3195	1.5126	1.7317	1.9799	2.2609	2.5785	2.9372	3.3417	3.7975
15	1.1610	1.2502	1.3459	1.5580	1.8009	2.0789	2.3966	2.7590	3.1722	3.6425	4.1772
16	1.1726	1.2690	1.3728	1.6047	1.8730	2.1829	2.5404	2.9522	3.4259	3.9703	4.5950
17	1.1843	1.2880	1.4002	1.6528	1.9479	2.2920	2.6928	3.1588	3.7000	4.3276	5.0545
18	1.1961	1.3073	1.4282	1.7024	2.0258	2.4066	2.8543	3.3799	3.9960	4.7171	5.5599
19	1.2081	1.3270	1.4568	1.7535	2.1068	2.5270	3.0256	3.6165	4.3157	5.1417	6.1159
20	1.2202	1.3469	1.4859	1.8061	2.1911	2.6533	3.2071	3.8697	4.6610	5.6044	6.7275
21	1.2324	1.3671	1.5157	1.8603	2.2788	2.7860	3.3996	4.1406	5.0338	6.1088	7.4002
22	1.2447	1.3876	1.5460	1.9161	2.3699	2.9253	3.6035	4.4304	5.4365	6.6586	8.1403
23	1.2572	1.4084	1.5769	1.9736	2.4647	3.0715	3.8197	4.7405	5.8715	7.2579	8.9543
24	1.2697	1.4295	1.6084	2.0328	2.5633	3.2251	4.0489	5.0724	6.3412	7.9111	9.8497
25	1.2824	1.4510	1.6406	2.0938	2.6658	3.3864	4.2919	5.4274	6.8485	8.6231	10.8347
26	1.2953	1.4727	1.6734	2.1566	2.7725	3.5557	4.5494	5.8074	7.3964	9.3992	11.9182
27	1.3082	1.4948	1.7069	2.2213	2.8834	3.7335	4.8223	6.2139	7.9881	10.2451	13.1100
28	1.3213	1.5172	1.7410	2.2879	2.9987	3.9201	5.1117	6.6488	8.6271	11.1672	14.4210
29	1.3345	1.5400	1.7758	2.3566	3.1187	4.1161	5.4184	7.1143	9.3173	12.1722	15.8631
30	1.3478	1.5631	1.8114	2.4273	3.2434	4.3219	5.7435	7.6123	10.0627	13.2677	17.4494

EXAMPLE 1 **Use Table 4.53 to find the compound amount and the compound interest on $6,000 at 10% compounded semiannually for 5 years.**

$$\text{Rate} = \frac{10\%}{2} = 5\%$$

$$\text{Periods} = 5 \times 2 = 10$$

$$\text{Factor} = 5\%,\ 10 \text{ periods} \Rightarrow 1.6289$$

Answers: Compound amount = 1.6289 × $6,000 = $9,773.40

Compound interest = $9,773.40 – $6,000 = $3,773.40

PRACTICE 1 **Use Table 4.53 to find the compound amount and the compound interest on $2,500 at 12% compounded quarterly for 4 years.**

Answers: Compound interest = $1,511.75; Compound amount = $4,011.75

EXAMPLE 2 **Roger deposits $1,700 in an account earning 8% interest compounded quarterly. How much money will he have at the end of three years?**

$$\text{Rate} = \frac{8\%}{4} = 2\%$$

Periods = 4 × 3 = 12

Factor = 2%, 12 periods ⇒ 1.2682 (from Table 4.53)

Answer: Compound amount = 1.2682 × $1,700 = $2,155.94

PRACTICE 2 **Fred deposits $3,000 in an account earning 6% interest compounded semiannually. How much money will he have at the end of 12 years?**

Answers: Compound interest = $3,098.40; Compound amount = $6,098.40

UNIT **53** Name ______________________ Date ______________

Compound Interest

Complete the following using Table 4.53:

Time	Principal	Rate	Compounded	Interest Rate per Compounding Period	# of Periods	Table Factor	Compound Amount	Compound Interest
3 yrs.	$2,000	8%	semiannually	4%	6	1.2653	$2,530.60	$530.60
2 yrs.	$1,500	4%	quarterly	1%	8	1.0829	$1,624.35	$124.35
4 yrs.	$2,500	6%	annually	6%	4	1.2625	$3,156.25	$656.25
5 yrs.	$800	6%	quarterly	1.5%	20	1.3469	$1,077.52	$277.52
6 yrs.	$5,000	8%	semiannually	4%	12	1.6010	$8,005.00	$3,005.00
8 yrs.	$3,000	10%	semiannually	5%	16	2.1829	$6,548.70	$3,548.70
10 yrs.	$2,500	7%	annually	7%	10	1.9672	$4,918.00	$2,418.00
6 yrs.	$1,800	12%	quarterly	3%	24	2.0328	$3,659.04	$1,859.04
4 yrs.	$4,000	8%	quarterly	2%	16	1.3728	$5,491.20	$1,491.20
7 yrs.	$6,000	6%	quarterly	1.5%	28	1.5172	$9,103.20	$3,103.20
3 yrs.	$2,000	12%	semiannually	6%	6	1.4185	$2,837.00	$837.00
10 yrs.	$15,000	14%	semiannually	7%	20	3.8697	$58,045.50	$43,045.50
12 yrs.	$9,000	8%	annually	8%	12	2.5182	$22,663.80	$13,663.80
10 yrs.	$2,000	10%	semiannually	5%	20	2.6533	$5,306.60	$3,306.60
7 yrs.	$3,500	12%	quarterly	3%	28	2.2879	$8,007.65	$4,507.65

Solve:

1. Carl Hasden lends $6,000 to the owner of a new bakery. He will be repaid at the end of 4 years at 6% compounded seminannually. How much will he be repaid?

   $7,600.80

2. Aaron invests $1,500 into an account earning 8% interest compounded quarterly. How much money will he have at the end of 5 years?

   $2,228.85

3. Nataja deposits $25,100 in Ranch Bank. Ranch pays 10% interest compounded semiannually. How much will Nataja have in her account at the end of 3 years?

   $33,636.51

4. The owner of a sporting goods store loaned $14,000 to Mel Cross to help him open an art shop. Mel plans to repay the owner at the end of 4 years with 12% interest compounded semiannually. How much will the owner receive at the end of four years?

   $22,313.20

5. Larry deposited $5,000 at Community Savings Bank at 9% interest compounded annually. What will Larry's deposit amount to after 6 years?

   $8,385.50

6. Ana and Philippe loan $8,000 to Rusty to open a music store. Rusty plans to repay Ana and Philippe at the end of 6 years with 8% interest compounded quarterly. How much will Ana and Philippe receive at the end of 6 years?

   $12,867.20

7. Rochelle needs $48,000 5 years from now to attend the college of her choice. If Rochelle's parents deposit $30,000 in a bank account paying 10% interest compounded semiannually, will they have the necessary amount at the end of 5 years?

   yes; $48,867

8. Monique wants to buy a new camper in 6 years. She estimates the camper will cost $6,400. Assume Monique invests $3,800 now at 8% interest compounded semiannually. Will Monique have enough money to buy the camper at the end of 6 years?

   no; $6,083.80

9. Gert deposits $8,240 in a bank that pays 12% compounded quarterly. Find the amount she will have at the end of 5 years.

   $14,882.26

10. What will the balance be in an account at the end of 6 years if the original deposit was $6,500, and the account earned 8% compounded semiannually.

    $10,406.50

CHAPTER 4

Name ____________________ Date ____________

Test

Answers

1. Change to a percent. (Round to the nearest tenth percent when necessary.)

   a. .53   b. .006   c. $\frac{1}{8}$ d. $\frac{7}{9}$

1a. 53%

1b. .6%

1c. 12.5%

1d. 77.8%

2. Change to a decimal.

   a. 18%   b. 4.14%

2a. .18

2b. .0414

3. Change to a fraction. (Reduce to lowest terms if necessary.)

   a. 37%   b. $24\frac{2}{3}\%$

3a. $\frac{37}{100}$

3b. $\frac{37}{150}$

4. Identify the base, portion, and percent.
   65 is 26% of 250.

4. Base = 250

   Portion = 65

   Rate = 26%

5. Solve: What percent of 150 is 60?

5. 40%

6. Solve: 65% of what number is 13?

6. 20

7. Solve: 66% of 90 is what number?

7. 59.4

8. This month credit customers paid $44,000 to a medical doctor. This amoung represents 20% of what all his customers owe. What is the total amount owed to the doctor?

8. $220,000

(Test continues on next page)

Answers

Test *(Concluded)*

9. The price of a microwave oven increased from $600 to $800. What was the percent of increase? (Round to the nearest tenth percent.)

9. 33.3%

10. Robin earns $600 per week plus 3% of sales over $6,500. Robin's sales were $14,000. How much did Robin earn?

10. $825

11. A computer lists for $1,200 with a trade discount of 22%. What is the computer's net price?

11. $936

12. A print shop received a $4,000 invoice dated May 12. Terms were 3/10, 1/15, n/30. On May 24, full payment was sent. What is the amount of discount?

12. $40

13. Kate and Mike borrow $30,000 at 12% for 5 years to finance a new sailboat. Calculate the simple interest on this loan.

13. $18,000

14. Fill in the blanks.

Principal	Time	Compound Interest Rate	Compounded
$200	1 year	8%	Quarterly

a. Number of compound periods: 14a. 4 periods

b. Rate for each period: 14b. 2%

c. Total interest (Note: Round *each* interest amount to the nearest cent): 14c. $16.48

CHAPTER

5 Introduction to Algebra

UNITS

UNIT

54 Signed Numbers

Up to this point, all the numbers we have used have been 0 or greater. In this chapter, beginning with this unit, we will expand our use of numbers to include numbers less than zero.

Numbers greater than zero are known as **positive numbers.** Numbers less than zero are known as **negative numbers.** Zero is neither positive nor negative. When we talk about **signed numbers** we are referring to zero, positive numbers, and negative numbers. The illustration below describes this signed number system.

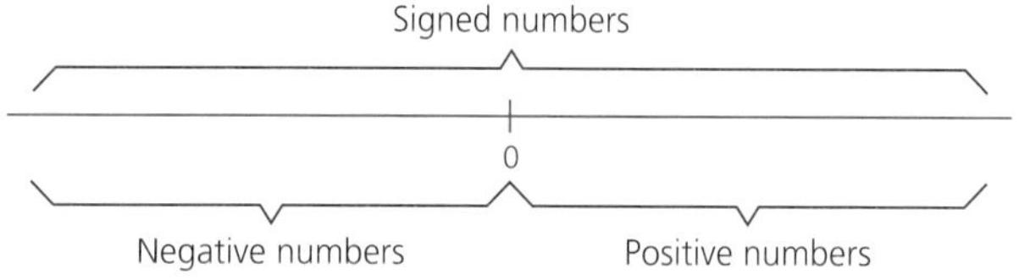

Positive numbers can be written with a positive symbol. For example, +7, +12, and $+\frac{1}{4}$ are all expressions of positive numbers and are read as "positive seven," "positive twelve," and "positive one-fourth." However, it is not necessary to write positive numbers using the positive sign, +. Any number without this positive symbol is understood to be positive. So 7, 12, and $\frac{1}{4}$ are still referred to as "positive seven," "positive twelve," and "positive one-fourth."

Negative numbers *must* be written with a minus sign preceding the number. For example, –5, –18, and $-\frac{3}{7}$ are all expressions of negative numbers and are read as "negative five," "negative eighteen," and "negative three-sevenths."

One way to familiarize oneself with the meaning of signed numbers is with a number line. If we were to graph the numbers –7, –5, –2, 4, and 9 on a number line they would appear as follows:

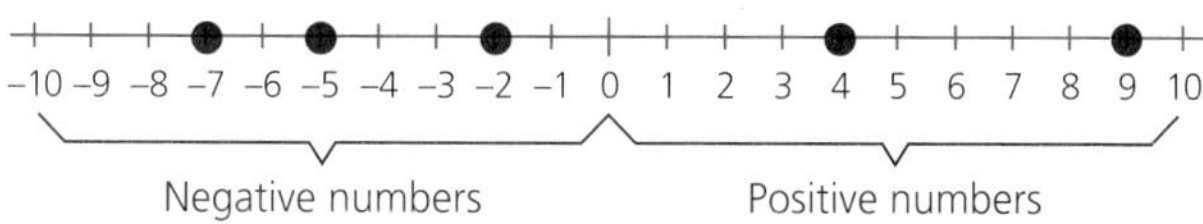

When comparing the relative size of signed numbers, we should remember that the smaller number always appears to the left of the other number; consequently, we can regard –7 as being smaller or less than –5.

Another useful point to remember is that the negative sign, –, can also mean *the opposite of*. The number –(–8) can be read as "the opposite of negative 8," which means 8 ("positive eight").

EXAMPLE 1 **On the number line provided, graph the following signed numbers.**

–4, –1, 0, 5, 7

Answer:

–10 –9 –8 –7 –6 –5 –4 –3 –2 –1 0 1 2 3 4 5 6 7 8 9 10

PRACTICE 1 **On the number line provided, graph the following signed numbers.**

–2, 5, 1, –6, 4, –5

–6 –5 –2 0 1 4 5

Answer:

–10 –9 –8 –7 –6 –5 –4 –3 –2 –1 0 1 2 3 4 5 6 7 8 9 10

EXAMPLE 2 **Identify the following numbers as positive or negative.**

Answers: 3 Positive –14 Negative +7 Positive $-\frac{2}{3}$ Negative

PRACTICE 2 **Identify the following numbers as positive or negative.**

Answers: –1 Negative –13 Negative +5 Positive $\frac{1}{2}$ Positive

EXAMPLE 3 **State the opposite of each of the following numbers.**

Answers: 2 –2 –16 16 $-\frac{1}{2}$ $\frac{1}{2}$ 4.7 –4.7

PRACTICE 3 **State the opposite of each of the following numbers.**

Answers: –14 14 $\frac{7}{8}$ $-\frac{7}{8}$ –5.2 5.2

UNIT **54** Name ______________________ Date __________

Signed Numbers

Graph the following numbers on the number lines provided:

1. –9, 2, 0, –3

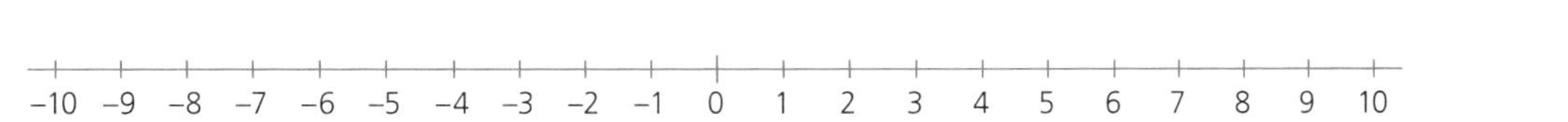

–10 –9 –8 –7 –6 –5 –4 –3 –2 –1 0 1 2 3 4 5 6 7 8 9 10

2. 5, 3, –4, –7

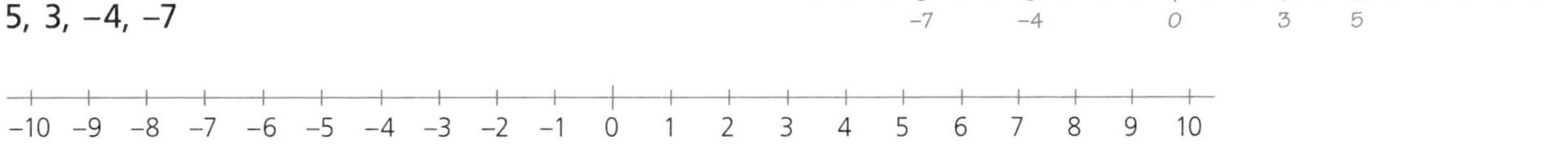

–10 –9 –8 –7 –6 –5 –4 –3 –2 –1 0 1 2 3 4 5 6 7 8 9 10

3. –7, 9, 4, –3, 0

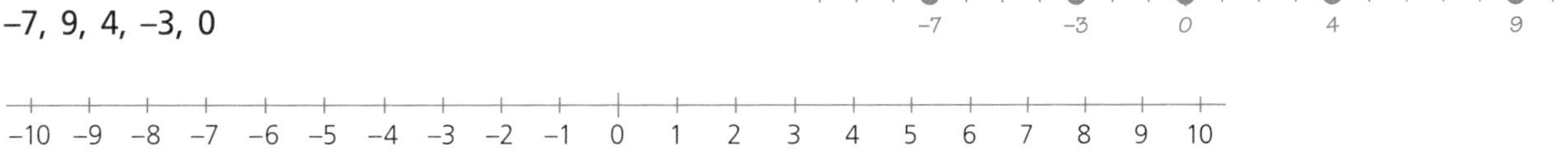

–10 –9 –8 –7 –6 –5 –4 –3 –2 –1 0 1 2 3 4 5 6 7 8 9 10

4. $-\frac{1}{2}, -1, 5, \frac{1}{2}$

−6 −5 −4 −3 −2 −1 0 1 2 3 4 5 6

5. $2, -\frac{3}{2}, -4, 6, 2\frac{1}{2}$

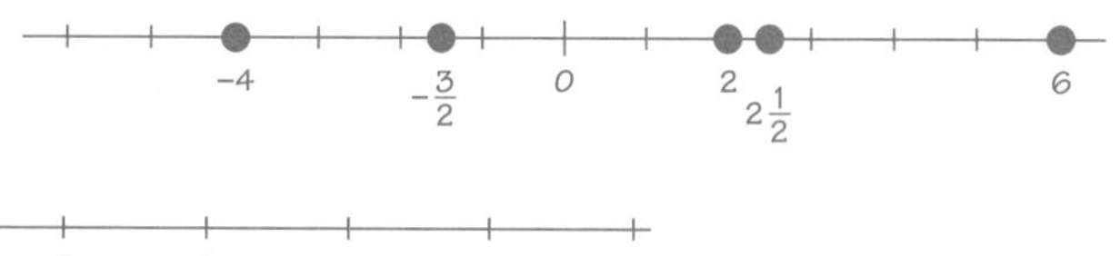

−6 −5 −4 −3 −2 −1 0 1 2 3 4 5 6

Identify the following numbers as positive or negative:

6. $-\frac{2}{3}$ Negative
7. 15 Positive
8. −5 Negative
9. 16 Positive
10. +3 Positive
11. $+\frac{7}{8}$ Positive
12. $-2\frac{3}{4}$ Negative
13. −27 Negative
14. −5.1 Negative
15. $16\frac{2}{3}$ Positive

State the opposite of each of the following numbers:

16. −32 32
17. −7 7
18. 3.1 −3.1
19. +43 −43
20. −22.2 22.2
21. $-\frac{4}{3}$ $\frac{4}{3}$
22. $16\frac{1}{2}$ $-16\frac{1}{2}$
23. 43.6 −43.6
24. 30 −30
25. −12 12

UNIT

55 Adding and Subtracting Signed Numbers

Adding signed numbers is accomplished by following these rules:

1. Add two or more signed numbers that have the same sign by adding the numbers themselves and attaching their common sign to the sum.
2. Add two signed numbers that have different signs by subtracting the smaller number from the larger without regard to their signs. Attach the sign of the larger number to the answer.

Subtracting two signed numbers is accomplished by adding the first number and the opposite of the second number. Try to remember this rule by reminding yourself that subtraction is the opposite operation of addition. Furthermore, the subtraction symbol, like the negative sign, can also mean *the opposite of*. Follow these rules to subtract signed numbers.

1. Change the sign of the number following the subtraction symbol to its signed opposite.
2. Change the subtraction symbol to an addition symbol.
3. Add the two signed numbers using the previously established rules for adding signed numbers.

EXAMPLE 1 **Add: (+4) + (+7).**

Answer: (+4) + (+7) = +11

Same signs, so add and keep their common sign.

PRACTICE 1 **Add: (−14) + (−12).**

Answer: −26

EXAMPLE 2 **Add: (−12) + (7).**

Answer: (−12) + (7) = −5

Different signs, so subtract and give the sign of the larger number to the answer.

PRACTICE 2 **Add: (7) + (−15).**

Answer: −8

EXAMPLE 3 **Subtract: (−6) − (−19).**

(−6) − (−19) = (−6) + (19) — Change −19 to 19 and change the subtraction symbol to addition.

Answer: (−6) + (19) = 13 — Different signs, so subtract and keep the sign of the larger number.

PRACTICE 3 **Subtract: (−18) − (32).**

Answer: −50

EXAMPLE 4 **Combine: (+7) + (−3) − (−12) − (6) + 22 − (−5).**

Change the signs of the numbers following the subtraction symbols to their signed opposite, and change all subtraction symbols to addition.

(+7) + (−3) + (12) + (−6) + 22 + (5)

Group numbers with like signs and add according to the rules for addition.

Answer: (+7) + (12) + 22 + (5) + (−3) + (−6) = 46 + (−9) = 37

PRACTICE 4 **Combine: (−4) − (−3) + (17) − (−12) + 8 − (−5).**

Answer: 41

UNIT **55** Name ____________________ Date ____________

Adding and Subtracting Signed Numbers

Add:

1. (−7) + (−3) =
   −10
2. 17 + (−5) =
   12
3. −21 + (−16) =
   −37
4. (−20) + (−25) =
   −45
5. (11) + (−8) =
   3
6. 13 + (−5) =
   8
7. (12) + (21) =
   33
8. (−29) + (−6) =
   −35
9. 14 + (−14) =
   0
10. 56 + (−13) =
    43
11. (−48) + (25)
    −23
12. −20 + 25 =
    5
13. 14 + (−32) =
    −18
14. −9 + (16) =
    7
15. 31 + (−83) =
    −52
16. (−24) + (56) =
    32
17. (−65) + (27) =
    −38
18. (−13) + (−18) =
    −31
19. 11 + (−13) =
    −2
20. −17 + (−25) =
    −42

UNIT **55** Name ______________________ Date ______________

Adding and Subtracting Signed Numbers

Subtract:

21. 6 – (–2) = 8
22. 4 – (–10) = 14
23. –9 – (–6) = –3
24. 8 – (2) = 6
25. 7 – (15) = –8
26. 14 – (20) = –6
27. 8 – (–6) = 14
28. 33 – (–27) = 60
29. –6 – (–7) = 1
30. 17 – (–3) = 20
31. –6 – (–6) 0
32. 8 – (–9) = 17
33. 19 – (–27) = 46
34. 6 – (20) = –14
35. 34 – (–8) = 42
36. –57 – (–79) = 22
37. 23 – (–10) = 33
38. 67 – (76) = –9
39. –68 – (90) = –158
40. –12 – (12) = –24

Business Applications:

41. A checking account with a balance of $63 is overdrawn because a check for $97 has been written on the account. Combine $63 and (–$97) to find the account's negative balance.

–$34

42. A checking account shows a balance of –$16. The owner deposits $50 into the account. Add these numbers to find the account's balance.

$34

Combine:

43. 5 – (–10) + 8 = 23

44. –10 – (–3) + 6 – (–5) = 4

45. $6 - (4) - (-3) + (12) =$ 17

46. $2 + (-4) - (-2) + (16) =$ 16

47. $-3 + (-2) - (-6) - (12) - (8) =$ −19

48. $14 - (-3) - (7) + (-22) + (-5) =$ −17

49. $-16 - (-12) + 18 - (-32) - 14 + (-57) =$ −25

50. $16 - (-12) + (-20) - (-25) - (-2) + 17 =$ 52

51. $2 - (-20) + (-7) - 18 + 9 - (-13) - (2) =$ 17

52. $-7 + (-5) + 14 - (-32) - (64) + (-13) - (-8) =$ −35

UNIT

56 Multiplication and Division of Signed Numbers

Signed numbers are easily multiplied by following these rules:

1. When multiplying two numbers with the same signs, the product is positive.
2. When multiplying two numbers with different signs, the product is negative.
3. When multiplying an odd number of negative numbers, the product is always negative.
4. When multiplying an even number of negative numbers, the product is always positive.

The division of signed numbers follows the exact same rules as the ones for multiplication of signed numbers.

1. When dividing two numbers with the same sign, the quotient is positive.
2. When dividing two numbers with different signs, the quotient is negative.

EXAMPLE 1 **Multiply the following.**

Answers: $(8) \times (-4) \Rightarrow$ different signs, product is negative = –32

$(-8) \times (-3) \Rightarrow$ same signs, product is positive = 24

PRACTICE 1 **Multiply the following.**

Answers: $(-3) \times (-15) =$ 45 $(-6) \times (7) =$ –42

EXAMPLE 2 **Divide the following.**

Answers: $\frac{-18}{-6} \Rightarrow$ same signs, quotient is positive = 3

$\frac{-32}{16} \Rightarrow$ different signs, quotient is negative = –2

$(-\frac{1}{3}) \div (\frac{2}{6}) = (-\frac{1}{3}) \times (\frac{6}{2}) = -\frac{6}{6} = -1$

PRACTICE 2 **Divide the following.**

Answers: $\frac{22}{11} =$ 2 $-\frac{24}{6} =$ –4 $(-\frac{3}{12}) \div (-\frac{1}{4}) =$ 1

EXAMPLE 3 **Multiply the following**

$(-3) \times (-2) \times (-5) \times (4) =$

Answer: $(-3) \times (-2) \times (-5) \times (+4) = (+6) \times (-5) \times (+4) = (-30) \times (+4) = -120$

Answer: $(-\frac{1}{3}) \times (\frac{6}{9}) \times (-81) = (-\frac{2}{9}) \times (-81) = 18$

PRACTICE 3 **Multiply the following.**

Answers: $(-2) \times (5) \times (-6) \times (-8) =$ -480 $(-\frac{4}{15}) \times (-\frac{5}{8}) \times (-\frac{2}{7}) =$ $-\frac{1}{21}$

UNIT **56** Name ______ Date ______

Multiplication and Division of Signed Numbers

Multiply:

1. $(7) \times (-6) =$ -42
2. $(-4) \times -32 =$ 128
3. $(-9) \times (-9) =$ 81
4. $(15) \times (-3) =$ -45
5. $(-13) \times (9) =$ -117
6. $(-12) \times (-4) =$ 48
7. $(14) \times (-50) =$ -700
8. $(-9) \times (-7) =$ 63
9. $(-\frac{4}{3}) \times (\frac{1}{2}) =$ $-\frac{2}{3}$
10. $(-.01) \times (-.4) =$ $.004$
11. $(-\frac{1}{6}) \times (-\frac{2}{8}) =$ $\frac{1}{24}$
12. $(15) \times (-\frac{2}{3}) =$ -10
13. $(-4.2) \times (-1.5) =$ 6.3
14. $(-.0032) \times (-.12) =$ $.000384$
15. $(34.1) \times (-.02) =$ -0.682
16. $(-9) \times (-6) \times (-8) =$ -432
17. $(27) \times (-3) \times (-4) =$ 324
18. $(-8) \times (-5) \times (-3) =$ -120
19. $(-7) \times (9) \times (-6) \times (4) =$ $1{,}512$
20. $(-\frac{1}{3}) \times (-\frac{6}{8}) \times (\frac{1}{2}) =$ $\frac{1}{8}$
21. $(-\frac{3}{4}) \times (-\frac{2}{12}) \times (\frac{1}{3}) =$ $\frac{1}{24}$

UNIT **56** Name ______________________ Date ____________

Multiplication and Division of Signed Numbers

22. $(-8) \times (\frac{3}{4}) \times (\frac{7}{9}) =$

    $-4\frac{2}{3}$

23. $(-10) \times (-10) \times (-10) =$

    $-1{,}000$

24. $(-\frac{27}{30}) \times (-\frac{1}{9}) \times (-60) =$

    -6

Divide:

25. $9 \div (-3) =$

    -3

26. $(-28) \div (-7) =$

    4

27. $(-48) \div (-6) =$

    8

28. $(-5) \div (5) =$

    -1

29. $(-72) \div (-8) =$

    9

30. $(-9) \div (3) =$

    -3

31. $(24) \div (-8) =$

    -3

32. $(20) \div (-5) =$

    -4

33. $(-120) \div (-10) =$

    12

34. $(-\frac{1}{2}) \div (\frac{1}{4}) =$

    -2

35. $(-1.2) \div (-2) =$

    0.6

36. $(-36) \div (-18) =$

    2

37. $(-\frac{4}{9}) \div (\frac{1}{3}) =$

    $-1\frac{1}{3}$

38. $(-28) \div (-\frac{4}{7}) =$

    49

39. $(-\frac{8}{9}) \div 8 =$

    $-\frac{1}{9}$

40. $(-0.23) \div (-0.23) =$

    1

41. $(-0.64) \div (-0.8) =$

    0.8

42. $(-0.56) \div (-0.07) =$

    8

43. $(39) \div (-0.03) =$

    $-1{,}300$

44. $(-\frac{5}{7}) \div (-5) =$

    $\frac{1}{7}$

45. $(-\frac{1}{3}) \div (-\frac{1}{9}) =$

    3

46. $(1\frac{1}{2}) \div (-\frac{2}{3}) =$

    $-2\frac{1}{4}$

47. $(-\frac{1}{11}) \div (-2) =$

    $\frac{1}{22}$

48. $(-\frac{3}{7}) \div (7) =$

    $-\frac{3}{49}$

UNIT 57 Order of Operations

Arithmetic problems that involve more than one operation must be completed by using a specific order of operations. This order of operations is:

1. Do all operations inside a set of parentheses.
2. Calculate all expressions that involve *exponents* or *roots.*
3. Multiply or divide from left to right.
4. Add or subtract from left to right.

An exponent is used to indicate repeated multiplication. For instance, in the expression 5^2, the **exponent** is the smaller number 2 and the number 5 is the **base.** This expression says to take two 5s and multiply them times themselves, so

$5^2 = 5 \times 5 = 25$

The number 5 can also be referred to as the **square root** of 25 or as the number that when multiplied times itself gives 25. We indicate this square root property by the expression $\sqrt{25}$.

EXAMPLE 1 **Simplify the following expressions.**

Answers: $2^3 = 2 \times 2 \times 2 = 8$

$4^2 = 4 \times 4 = 16$

$(-7)^2 = (-7) \times (-7) = 49$

$\sqrt{16} \Rightarrow$ What times itself gives 16? = 4

$\sqrt{64} \Rightarrow$ What times itself gives 64? = 8

PRACTICE 1 **Simplify the following expressions.**

Answers: $4^3 =$ 64 $\quad (-9)^2 =$ 81 $\quad \sqrt{100} =$ 10 $\quad \sqrt{36} =$ 6

EXAMPLE 2 **Use the order of operations to solve the following problem.**

$16 + 2^3 \times (8 + 3) \div 4$ $\Rightarrow$ Do operation inside grouping symbols first.

$= 16 + 2^3 \times (11) \div 4$ $\Rightarrow$ Calculate exponent.

$= 16 + 8 \times (11) \div 4$ $\Rightarrow$ Multiply.

$= 16 + 88 \div 4$ $\Rightarrow$ Divide.

$= 16 + 22$ $\Rightarrow$ Add.

Answer: $= 38$

PRACTICE 2 **Use the order of operations to solve the following problem.**

$13 + 6^2 \div (2 + 7) \times 8 =$

Answer: 45

UNIT **57** Name ____________________ Date ____________

Order of Operations

Simplify each of the following expressions:

1. $\sqrt{81} =$ 9
2. $\sqrt{144} =$ 12
3. $3^3 =$ 27
4. $2^2 =$ 4
5. $5^3 =$ 125
6. $14^2 =$ 196
7. $7^2 =$ 49
8. $10^3 =$ 1,000
9. $\sqrt{49} =$ 7
10. $\sqrt{225} =$ 15
11. $\sqrt{121} =$ 11
12. $\sqrt{36} =$ 6
13. $20^2 =$ 400
14. $12^2 =$ 144
15. $(-3)^2 =$ 9
16. $(-13)^2 =$ 169
17. $(-3)^3 =$ −27
18. $\sqrt{289} =$ 17
19. $\sqrt{0} =$ 0
20. $\sqrt{64} =$ 8

Use the order of operations to solve the following problems:

21. $(6 - 2)^2 =$

    16

22. $(4 - 2)^2 =$

    4

23. $5 + (3 + 2)^2 =$

    30

24. $(9 - 3)^2 \div 6 \times 4 =$

    24

25. $2^2 - 7 \times (3 + 2)^2 =$

    −171

26. $5 + (13)^2 - 8 \times (6 + 2) =$

    110

27. $(-7)^2 - 49 =$

    0

28. $4 + 2 \times (7 + 9) =$

    36

29. $2 \times (6 - 4) =$

    4

30. $36 \div (3^2) + (-5) =$

    −1

31. $-2 + 4 \times (7 - 3) =$

    14

32. $(-49) \div (3 + 4) =$

    −7

33. $5 + (8 - 12)^3 =$

    −59

34. $5 \times (-2)^2 + 6 \div 2 =$

    23

35. $8 + 3^2 \times 4 - (6 + 4) =$

    34

UNIT

58 Writing and Evaluating Expressions

Letters are often used in algebra to symbolize quantities whose size can vary or whose size is unknown. In algebra these letters are called **variables.** Often, variables are used along with numbers or **constants** to express a relationship. For example, the statement "5 more than a number" can be symbolized by the mathematical statement $5 + n$ where 5 is the constant and n is the variable that represents the unknown quantity. The entire statement $5 + n$ is called an **expression.**

One important point to remember is that multiplication is indicated *without* the use of the multiplication sign, ×. If the multiplication sign were used, it might be confused with the variable x.

Expressing multiplication without the use of x is done in this fashion:

Multiplication of a constant and a variable is indicated by placing these quantities side by side with no symbol between them. For example, "3 times a number would be written as $3n$.

Multiplication of two constants is indicated in one of two ways. One approach is to write each constant inside a set of parentheses and place the parentheses side by side with no symbol between them. The other approach is to write one of the numbers inside a set of parentheses and the other to the left of the parentheses with no symbol between them. For example 4 times 8 can be written as (4)(8) or as 4(8).

Expressions composed of letters and numbers are evaluated by substituting specific values for the variables and by using the order of operations to find the expression's value.

EXAMPLE 1 **Write expressions for the following statements. Use *x* for the variable.**

Answers:

3 less than a number ⇒ $x - 3$
5 times a number ⇒ $5x$
4 less than the sum of a number and 2 ⇒ $(x + 2) - 4$

PRACTICE 1 **Write expressions for the following statements. Use *x* for the variable.**

Answers:

12 more than a number ⇒ $12 + x$

4 times a number ⇒ $4x$

1 less than the sum of a number and three ⇒ $(x + 3) - 1$

EXAMPLE 2 **Evaluate the following expression.**

$3x + 2y$ when $x = 2$ and $y = -4$

Answer: $3x + 2y = 3(2) + 2(-4) = 6 + (-8) = -2$

PRACTICE 2 **Evaluate the following expression:**

$-4x + 13y$ when $x = -4$ and $y = 2$

Answer: 42

UNIT 58 Name ______ Date ______

Writing and Evaluating Expressions

Write the expressions for the following statements. Use *x* for the variable.

1. The sum of a number and seven.
   $x + 7$
2. Eighteen less than a number.
   $x - 18$
3. Three times a number increased by four.
   $3x + 4$
4. Six times a number plus two.
   $6x + 2$
5. Six times the sum of a number plus two.
   $6(x + 2)$
6. The sum of a number and seven.
   $x + 7$
7. The product of six and three times a number.
   $6(3x)$
8. The product of two and a number, minus two.
   $2x - 2$

Evaluate the following expressions:

9. $3x - 8$ when $x = 10$ 22
10. $y^2 - 5y$ when $y = 7$ 14
11. $\frac{x + 5}{3}$ when $x = 10$ 5
12. $5x - 2x - 3$ when $x = 3$ 6
13. $\frac{6x - 15}{2x - 5}$ when $x = 8$ 3
14. $4(5x + 3) - x$ when $x = 4$ 88
15. $7(4 + 12x)$ when $x = 2$ 196
16. $3x - 2y + 6$ when $x = 4$ and $y = -2$ 22
17. $3x^2 - 12y$ when $x = 3$ and $y = 1$ 15
18. $30 - \frac{1}{6}(y^2 - y)$ when $y = 7$ 23
19. $5y - 2(y + 3)$ when $y = 4$ 6
20. $2(3x - 5) + (3y + 2)$ when $x = 6$ and $y = -3$ 19
21. $(4x - 2)^2 - 6y$ when $x = 4$ and $y = 3$ 178
22. $\frac{1}{2}xy - (x + y)^2$ when $x = 6$ and $y = -4$ −16
23. $(x + 3)(x + 5)$ when $x = -5$ 0

Business Applications:

24. Overtime pay is paid at the rate of two times the hourly wage. Write an expression for the rate of overtime pay.

    $2x$

25. The second shift processes four times as many orders as the first shift. Write an expression for the number of orders processed by the second shift.

    $4x$

UNIT

59 Using Addition and Subtraction to Solve Equations

An **equation** is an expression of equality between two quantities. An equals sign separates the two quantities thus creating a left-hand side and a right-hand side in the equation.

For example, the equation $x + 7 = 16$ expresses the equality of the quantities $x + 7$ and 16.

Solving an equation is the process of finding the value of the variable in the equation that when substituted for the variable will produce a true statement. Often this process can be done simply by inspection. For instance, in the equation $x + 7 = 16$, we are looking for a number that when added to 7 will produce 16. The number is 9. However, not all equations can be solved by inspection. Consequently, we try to think of an equation as a balancing scale where the equals sign is the center balance.

To keep the equation "in balance," whatever we mathematically do to one side must be done to the other side also. An important rule to remember is this:

> **The same number can be added or subtracted to *both* sides of an equation without destroying the equality.**

Solving an equation means that we must somehow isolate the variable on one side of the equals sign with the constant isolated on the other side.

To do this, examine the equation.

1 If the equation is of the form

$x + \mathrm{k} = \mathrm{c}$ or $\mathrm{k} + x = \mathrm{c}$,

where k and c are constants, subtract k from both sides. Doing this operation will isolate the variable x on one side of the equals sign.

2 If the equation is of the form

$x - \mathrm{k} = \mathrm{c}$ or $-\mathrm{k} + x = \mathrm{c}$,

where k and c are constants, add k to both sides. Doing this operation will isolate the variable x on one side of the equals sign.

EXAMPLE 1 **Solve for *x*.**

$$x + 7 = 14 \Rightarrow \begin{array}{rcr} x + 7 & = & 14 \\ -7 & & -7 \\ \hline & x = & 7 \end{array}$$

The equation is of the form $x + \mathrm{k} = \mathrm{c}$, so subtract 7 from both sides.

Answer: $x = 7$

PRACTICE 1 **Solve for *x*.**

$x + 5 = 17$

Answer: x = 12

EXAMPLE 2 **Solve for *x*.**

$x - 12 = 21 \Rightarrow \begin{array}{rcr} x - 12 & = & 21 \\ +12 & & +12 \\ \hline & x = & 33 \end{array}$ The equation is of the form $x - k = c$, so add 12 to both sides.

Answer: $x = 33$

PRACTICE 2 **Solve for *x*.**

$x - 11 = 12$

Answer: $x = 23$

EXAMPLE 3 **Solve for *x*.**

$-7 + x = 5 + 9 \Rightarrow$ Combine 5 + 9.

$-7 + x = 14 \Rightarrow \begin{array}{rcr} -7 + x & = & 14 \\ +7 & & +7 \\ \hline & x = & 21 \end{array}$ The equation is of the form $-k + x = c$, so add 7 to both sides.

Answer: $x = 21$

PRACTICE 3 **Solve for *x*.**

$-4 + x = -6 + 12$

Answer: $x = 10$

EXAMPLE 4 **Solve for *x*.**

$-12 + x = 4 + (-16) \Rightarrow$ Combine 4 + (–16).

$-12 + x = -12 \Rightarrow \begin{array}{rcr} -12 + x & = & -12 \\ +12 & & +12 \\ \hline & x = & 0 \end{array}$ The equation is of the form $-k + x = c$, so add 12 to both sides.

Answer: $x = 0$

PRACTICE 4 **Solve for *x*.**

$7 + x = 9 + (-2)$

Answer: $x = 0$

UNIT **59** Name ______________________ Date ______________

Using Addition and Subtraction to Solve Equations

Solve the equations:

1. $x + 3 = 9$

   $x = 6$

2. $x - 3 = 5$

   $x = 8$

3. $x - 12 = 16$

   $x = 28$

4. $x + 9 = 12$

   $x = 3$

5. $11 + x = 26$

   $x = 15$

6. $x - 2 = 13$

   $x = 15$

7. $-6 + x = -17$

   $x = -11$

8. $-18 + x = 23$

   $x = 41$

9. $15 + x = 40$

   $x = 25$

10. $9 + x = -17$

    $x = -26$

11. $x - \frac{1}{2} = 2$

    $x = 2\frac{1}{2}$

12. $-10 + x = 6$

    $x = 16$

13. $6 = x - 8$

    $x = 14$

14. $12 = x + 3$

    $x = 9$

15. $11 = x - 7$

    $x = 18$

16. $x + 3 = 6 - 4$

    $x = -1$

17. $x - 5 = 12 + 8$

    $x = 25$

18. $x - \frac{3}{2} = -3$

    $x = -1\frac{1}{2}$

19. $-7 + x = 4$

    $x = 11$

20. $x - 8 = -4$

    $x = 4$

21. $-4 + x = -5$

    $x = -1$

22. $x + 1 = 9$

    $x = 8$

23. $x + \frac{2}{3} = 5\frac{2}{3}$

    $x = 5$

24. $x - 3.6 = 2.7$

    $x = 6.3$

25. $x - \frac{1}{2} = -4$

    $x = -3\frac{1}{2}$

26. $8 + x = -8$

    $x = -16$

27. $-3 = -7 + x$

    $x = 4$

28. $x - 24 = -72$

    $x = -48$

29. $-6 + 2 = x - 5$

    $x = 1$

30. $x + 4 = -1 + 15$

    $x = 10$

UNIT

60 Using Multiplication and Division to Solve Equations

Another important rule to remember when solving equations that involve multiplication and division is:

The same nonzero number can be multiplied or divided on both sides of an equation without destroying the equality.

The following steps outline the procedure used for solving equations that involve multiplication or division. Remember, multiplication is an understood operation in algebra.

1. If the equation is of the form $ax = c$, where a and c are constants, divide both sides of the equation by a.
2. If the equation is of the form $\frac{x}{a} = c$, where a and c are constants, multiply both sides of the equation by a.

EXAMPLE 1 **Solve for *x*.**

$3x = -27$

Answer: $\frac{3x}{3} = \frac{-27}{3} \Rightarrow x = -9$

The equation is of the form $ax = c$, so divide both sides by 3.

PRACTICE 1 **Solve for *x*.**

$7x = -56$

Answer: $x = -8$

EXAMPLE 2 **Solve for *x*.**

$\frac{x}{-5} = 10$

Answer: $-5\left(\frac{x}{-5}\right) = -5(10) \Rightarrow x = -50$

The equation is of the form $\frac{x}{a} = c$, so multiply both sides by -5.

PRACTICE 2 **Solve for *x*.**

$\frac{x}{-3} = 6$

Answer: $x = -18$

UNIT **60** Name ______________________ Date ______________

Using Multiplication and Division to Solve Equations

Solve for *x*:

1. $2x = 8$

   $x = 4$

2. $-3x = 18$

   $x = -6$

3. $-4x = -20$

   $x = 5$

4. $-5x = -40$

   $x = 8$

5. $6x = -36$

   $x = -6$

6. $-8x = 40$

   $x = -5$

7. $7x = 42$

   $x = 6$

8. $-32 = 8x$

   $x = -4$

9. $12x = -60$

   $x = -5$

10. $-6x = 90$

    $x = -15$

11. $14x = 28$

    $x = 2$

12. $-9x = -81$

    $x = 9$

13. $64 = -8x$

    $x = -8$

14. $-168 = -3x$

    $x = 56$

15. $-22 = 11x$

    $x = -2$

16. $\frac{x}{3} = 6$

    $x = 18$

17. $\frac{x}{2} = -15$

    $x = -30$

18. $\frac{x}{-4} = -3$

    $x = 12$

19. $\frac{x}{6} = -11$

    $x = -66$

20. $\frac{x}{-12} = -3$

    $x = 36$

21. $\frac{x}{-4} = -10$

    $x = 40$

22. $\frac{x}{3} = 15$

    $x = 45$

23. $\frac{x}{9} = -3$

    $x = -27$

24. $\frac{x}{-8} = -4$

    $x = 32$

25. $\frac{x}{-20} = 2$

    $x = -40$

26. $\frac{x}{13} = 3$

    $x = 39$

27. $\frac{x}{-9} = -9$

    $x = 81$

28. $\frac{x}{-7} = 6$

    $x = -42$

29. $\frac{x}{-5} = 12$

    $x = -60$

30. $\frac{x}{-15} = -4$

    $x = 60$

The Distributive Property and Combining Similar Terms

To be able to solve all types of equations one must be able to recognize and combine similar algebraic terms. In algebra only similar terms can be added or subtracted. Terms are similar if their variables are exactly alike.

For instance, the following are all similar terms:

$7x$ and $5x$
$12y$ and $27y$
$-2r$ and $13r$

To combine similar algebraic terms:

1. Add or subtract the number parts of the expressions, remembering the rules for signed numbers previously learned.

Another important procedure to be familiar with is distribution. **Distribution** is a way of expanding expressions by using multiplication.

For instance

$2(3 + 6)$

can be solved using an order of operations.

$2(3 + 6) = 2(9) = 18.$

If we desired, we could compute this same answer by multiplying each term inside the parentheses by the multiplier outside the parentheses.

$2(3 + 6) = 2 \times 3 + 2 \times 6 = 6 + 12 = 18.$

The above is an example of the distributive property.

To use the distributive property:

1. When multiplying numerical terms, multiply *each* term inside a set of parentheses by the multiplier outside the parentheses.
2. When multiplying algebraic terms and numbers, multiply only the number parts and keep the letter parts the same. If there is no number in front of the letter, it is understood to be the number 1.

EXAMPLE 1 **Combine the following.**

$3x - (-7x) + 4x =$

Answer: $3x + (7x) + 4x = 14x$

All terms are similar, so we can combine the number parts.

$3x + 4y =$

Answer: 3x + 4y

Cannot be combined. These are not similar terms.

$4x - 2y + 6x =$

Answer: $4x + 6x - 2y = 10x - 2y$

Only $4x$ and $6x$ can be combined.

PRACTICE 1 **Combine the following.**

Answers: $-3x - 8y + 4x =$ x − 8y $2x - (-4x) + 6x =$ 12x

EXAMPLE 2 **Use the distributive property to simplify the expression.**

$4(3 + 9) =$

Answer: $4(3 + 9) = 4 \times 3 + 4 \times 9 = 12 + 36 = 48$

PRACTICE 2 **Use the distributive property to simplify the expression.**

Answer: $7(2 + 4) =$ 42

EXAMPLE 3 **Use the distributive property to simplify the expression.**

$-2(x + 3y) =$

Answer: $-2(x + 3y) = -2x - 6y$

PRACTICE 3 **Use the distributive property to simplify the expression.**

$-3(x + 4y) =$ $-3x - 12y$

UNIT **60** Name ______________________ Date ______________

The Distributive Property and Combining Similar Terms

Combine these terms:

1. $8x - 3x + 2x =$

   $7x$

2. $-4y + 8y + 2y =$

   $6y$

3. $16x - 5x + (-2x) =$

   $9x$

4. $8a - 5a + 7a =$

   $10a$

5. $-7a + 6a + (-9a) =$

   $-10a$

6. $3x + 5x + 2y =$

   $8x + 2y$

7. $7y - 3y + 5y =$

   $9y$

8. $3a - (-5a) + 2 =$

   $8a + 2$

9. $3x + 6y + 8x + 2y =$

   $11x + 8y$

10. $3x - 2x - 3x =$

    $-2x$

11. $-17x + 14x - 20y =$

    $-3x - 20y$

12. $9y - (11y) + 5y =$

    $3y$

13. $-3b + 4a + 5b =$

    $4a + 2b$

14. $2x - (-6x) + 5y + (-3y) =$

    $8x + 2y$

15. $13x - (-17x) + 4y + 2x =$

    $32x + 4y$

UNIT **60** Name ______________________________ Date ______________

The Distributive Property and Combining Similar Terms

Use the distributive property to simplify the following expressions:

16. $3(8 + 4) =$

36

17. $5(7 + 3) =$

50

18. $-6(5 + 4) =$

-54

19. $4(5 - 2) =$

12

20. $3(-6 + 4) =$

-6

21. $7(-8 + 9) =$

7

22. $14(2 + 3) =$

70

23. $-3(x + y) =$

$-3x - 3y$

24. $-4(x + y - z) =$

$-4x - 4y + 4z$

25. $2(4x + y) =$

$8x + 2y$

26. $5(x - y + 2z) =$

$5x - 5y + 10z$

27. $3(x - 3y) =$

$3x - 9y$

28. $4(p - 20) =$

$4p - 80$

29. $-6(x + 5) =$

$-6x - 30$

30. $8(p + 11) =$

$8p + 88$

UNIT

62 Solving Equations with Similar Terms and Parentheses

Solving more complex equations requires the use of a step-by-step procedure that when followed will prevent errors and help one to arrive at the equation's solution more effectively.

1 Remove parentheses by using the distributive property.

2 Combine similar terms on each side of the equation.

3 Use addition or subtraction to move all algebraic terms to one side and all constants to the other side.

4 Multiply or divide both sides of the equation by the same number to solve the equation.

EXAMPLE 1 **Solve for *x*.**

$12x + 4 = 5x - 10$

$$\begin{array}{rcl} 12x + 4 & = & 5x - 10 \\ -4 & & -4 \\ \hline 12x & = & 5x - 14 \\ -5x & & -5x \\ \hline \frac{7x}{7} & = & \frac{-14}{7} \end{array}$$

1. No parentheses to remove.
2. No similar terms to combine on each side.
3. Start by subtracting terms to move algebraic terms to one side and constants to another.
4. Divide each side by 7.

Answer: $x = -2$

PRACTICE 1 **Solve for *x*.**

$5x + 6 = 3x - 8$

Answer: $x = -7$

EXAMPLE 2 **Solve for *x*:**

$3(x - 1) = 7 - 2x$

$3x - 3 = 7 - 2x$

$$\begin{array}{rcl} 3x - 3 & = & 7 - 2x \\ +3 & & +3 \\ \hline 3x & = & 10 - 2x \\ 2x & & +2x \\ \hline \frac{5x}{5} & = & \frac{10}{5} \end{array}$$

1. Remove parentheses.
2. Add terms to move algebraic terms to one side and constants to the other.
3. Divide each side by 5.

Answer: $x = 2$

PRACTICE 2 **Solve for x.**

$3x + 2 = 2(x + 1)$ **Answer:** $x = 0$

UNIT Name ________________ Date ________

62 *Solving Equations with Similar Terms and Parentheses*

Solve for x:

1. $2x - 5 = 11$

   $x = 8$

2. $3x - 6 = 21$

   $x = 9$

3. $5x - 24 = -x$

   $x = 4$

4. $4x - 42 = -3x$

   $x = 6$

5. $3x - 2 = 14 - x$

   $x = 4$

6. $5x - 11 = -3 - 3x$

   $x = 1$

7. $3x - 15 = -x + 21$

   $x = 9$

8. $12x - 7 = -8x - 2$

   $x = \frac{1}{4}$

9. $6x - 5 = 1 - 3x$

   $x = \frac{2}{3}$

10. $12x - 4 = 3 - 2x$

    $x = \frac{1}{2}$

11. $4x + 7 = 3x + 15$

    $x = 8$

12. $2(x - 4) = 12$

    $x = 10$

13. $9x - 7 = 5(2x - 3)$

    $x = 8$

14. $5(x - 7) = 25$

    $x = 12$

15. $6(x + 2) = 30$

    $x = 3$

16. $5(10 - 3x) = 7(7 - 2x)$

    $x = 1$

17. $2 + 3(x - 3) = 8$

    $x = 5$

18. $7 + 4(2x - 9) = 19$

    $x = 6$

19. $-2(x + 3) = 7 - x$

    $x = -13$

20. $2(2x + 3) = 3x + 4$

    $x = -2$

21. $8(2 + x) = 5x + 1$

    $x = -5$

22. $6x - 7 = 1 - 10x$

    $x = \frac{1}{2}$

23. $5x + 4 = 2x - 5$

    $x = -3$

24. $8x - 9 = 2(5x - 6)$

    $x = \frac{3}{2}$

25. $6x - 4 = 3(-1 + x)$

    $x = \frac{1}{3}$

26. $5x + 30 = 4 - 8x$

    $x = -2$

27. $15x + 13 = 2(1 - 9x)$

    $x = -\frac{1}{3}$

28. $5x - 7 = 20x - 1$

    $x = -\frac{2}{5}$

29. $12x - 3 = 21x - 3$

    $x = 0$

30. $5 + 3(3 - x) = 3x + 12$

    $x = \frac{1}{3}$

UNIT

63 How to Dissect and Solve Word Problems Using Equations

The use of equations to solve word problems involves the creation of algebraic expressions to represent the ideas presented in the application. The expressions are then used to form an equation, which is then solved.

Although there is no one way to create these expressions and form equations from them, there are some general guidelines.

1. Read and reread the entire problem to make sure the information presented is understood.
2. Determine what the problem is asking for and gather all facts needed for solving the problem.
3. Express the unknown quantity as a single variable. If there is more than one unknown, first select a quantity to be the single variable according to one of these criteria.
   a. The quantity that seems to be the focus of all comparisons in the problem.
   or
   b. The quantity that seems to be the smallest.

   Then, describe all other unknowns in terms of this single variable
4. Write an equation, using the variables and expressions formed, that accurately expresses the situation presented in the problem.
5. Solve the equation.
6. Check the solution and evaluate its reasonableness.

The following blueprint aid can be a helpful tool for organizing your thoughts, along with the information in the problem itself.

Unknown(s)	*Variable(s)*	*Relationship*

EXAMPLE 1 **A pair of Nike sneakers were reduced \$30. The sale price was \$70. What was the original price?**

Unknown(s)	*Variable(s)*	*Relationship*
Original price	p = Original price	$p - 30$ = Sale price

$$\begin{array}{rcl} p - \$30 & = & \$70 \\ +30 & & +30 \\ \hline p & = & \$100 \end{array}$$

Answer: $p = \$100$

PRACTICE 1 **A pair of ski boots were marked down \$40 at an end of season clearance sale. The sale price was \$35. What was the original price?**

Answer: \$75

EXAMPLE 2 **Together, Barry Sullivan and Mitch Ryan sold 300 homes for Regis Realty. Barry sold nine times as many homes as Mitch. How many did each sell?**

Unknown(s)	*Variable(s)*	*Relationship*
# Homes sold by Barry Sullivan	$9h$	$9h + h = 300$ Homes
# Homes sold by Mitch Ryan	h	

$$9h + h = 300 \Rightarrow 10h = 300 \Rightarrow \frac{10h}{10} = \frac{300}{10} \Rightarrow h = 30$$

Answer: Mitch Ryan = h = 30 homes
Barry Sullivan = $9h = 9(30) = 270$ homes

PRACTICE 2 **Diane and René together transported a total of 12 children to a Cub Scout pack meeting in their 2 cars. Diane transported twice as many children in her van as René did in her station wagon. How many did each transport?**

Answer: René transported four children; Diane transported eight children

EXAMPLE 3 **Andy sold watches for $9 and alarm clocks for $5 at a flea market. Total sales were $287. People bought four times as many watches as alarm clocks. How many of each did Andy sell?**

Unknown(s)	*Variable(s)*	*Price*	*Relationship*
# Watches sold	4c	$9	$9(4c) + 5(c) = 287$
# Alarm clocks sold	c	$5	

$$9(4c) + 5(c) = 287 \Rightarrow 36c + 5c = 287 \Rightarrow \frac{41c}{41} = \frac{287}{41} \Rightarrow c = 7$$

Answer: Alarm clocks = $c = 7$
Watches = $4c = 4(7) = 28$

PRACTICE 3 **Sheffield Corporation sells sets of pots for $14 and a set of dishes for $12. On Labor Day weekend at a local charity, Sheffield's total sales were $1,080. People bought three times as many pots as dishes. How many of each set did Sheffield sell?**

Answer: 20 sets of dishes; 60 sets of pots

UNIT **63** Name ______________________ Date ______________

How to Dissect and Solve Word Problems Using Equations

Solve the following problems:

1. A Liz Claiborne sweater was reduced $40. The sale price was $85. What was the original price?

   $125

2. Kelly Doyle budgets $\frac{1}{8}$ of her yearly salary for entertainment. Kelly's total entertainment bill for the year is $6,500. What is Kelly's yearly salary?

   $52,000

3. Micro Knowledge sells 5 times as many computers as does Morse Electronics. The difference in sales between the two stores is 20 computers. How many computers did each store sell?

   Micro = 25 computers; Morse = 5 computers

4. Susie and Cara sell stoves at Elliott's Appliance. Together they sold 180 stoves in January. Susie sold five times as many stoves as Cara. How many stoves did each sell?

   Susie = 150 stoves; Cara = 30 stoves

5. Pasquale's Pizza sells meatball pizzas for $6 and cheese pizzas for $5. In March, Pasquale's total sales were $1,600. People bought twice as many cheese pizzas as meatball pizzas. How many of each did Pasquale sell?

   100 meatball pizzas; 200 cheese pizzas

6. U.S. Air reduced its airfare to California by $60. The sale price was $95. What was the original price?

   $155

7. Blue Furniture Company ordered sleepers for $300 each and nonsleepers for $200 each for a total cost of $8,000. Blue expects sleepers to outsell nonsleepers two to one. How many units of each were ordered?

   20 sleepers; 10 nonsleepers

8. American Airlines reduced its round-trip ticket from Boston to Chicago by $56. The sale price was $299. What was the original price?

   $355

9. Molly's Diner receives cash for $\frac{1}{7}$ of its sales. During the first week in September, Molly's cash sales were $490. What were Molly's total sales?

   $3,430

10. Soo Lin and Hubert Krona sell cars for Northland Auto. Over the past year, they sold 150 cars. Soo sells four times as many cars as Hubert. How many cars did each sell?

    Soo = 120 cars; Hubert = 30 cars

11. Nanda Yueh and Lane Zuriff sell homes for Margate Realty. Over the past 6 months, they sold 120 homes. Nanda sold three times as many homes as did Lane. How many homes did each sell?

Nanda = 90 homes; Lane = 30 homes

12. Runyon Company sells T-shirts for $2 and shorts for $4. In April, total sales were $600. People bought four times as many T-shirts as shorts. How many T-shirts and shorts did Runyon sell?

50 shorts; 200 T-shirts

13. Sears reduced its price on lawn mowers by $45. The sale price was $169. What was the original price?

$214

14. Marge budgets $\frac{1}{4}$ of her yearly salary for clothing. Marge's total clothing bill for the year is $15,000. What is her yearly salary?

$60,000

15. Bill's Roast Beef sells five times as many sandwiches as Pete's Deli sells. The difference between their sales is 360 sandwiches. How many sandwiches did each sell?

Bill's = 450 sandwiches; Pete's = 90 sandwiches

16. In February, Shelly sold four times as many boats as Rusty sold. The difference between their sales is 21 boats. How many boats did each sell?

Rusty = 7 boats; Shelly = 28 boats

17. The Computer Store sells diskettes for $3 each and small boxes of computer paper for $5 each. In August, total sales were $960. Customers bought five times as many diskettes as boxes of computer paper. How many of each did the Computer Store sell?

240 diskettes; 48 boxes of paper

18. At Maplewood Marine, shift 1 produced three times as much as shift 2 did. Maplewood's total production for June was 6,400 rowboats. What was the output for each shift?

Shift 1 = 4,800 rowboats; Shift 2 = 1,600 rowboats

19. Jarvis Company sells thermometers for $2 each and hot water bottles for $6 each. In December, Jarvis's total sales were $1,200. Customers bought seven times as many thermometers as hot water bottles. How many of each did Jarvis sell?

420 thermometers; 60 hot water bottles

20. Amtrack reduced its round-trip ticket from Boston to Washington by $49. The sale price was $140. What was the original price?

$189

CHAPTER 5

Name ______________________ Date __________

Test

Answers

1. Graph the following numbers on the number line provided:
   −6, −1, 0, 5, 7

   −6 −1 0 5 7

   1. 

2. Identify the following numbers as positive or negative.
   a. −3 b. −3.2 c. 5.7 d. +4.3

   2a. negative
   2b. negative
   2c. positive
   2d. positive

3. State the opposite of each of the following numbers.
   a. −12 b. 16 c. +27.6 d. $-\frac{2}{3}$

   3a. 12
   3b. −16
   3c. −27.6
   3d. $\frac{2}{3}$

4. Add: (−32) + (−17) + (22) = 4. −27

5. Subtract: (−13) − (−2) = 5. −11

6. Combine: (−3) − (−12) + (24) + (−8) − (6) = 6. 19

7. Multiply: (−5) × (3) × (−7) = 7. 105

(Test continues on next page)

Test (Concluded)

Answers

8. Divide: $(-\frac{2}{7}) \div (\frac{3}{14}) =$ — 8. $-\frac{4}{3}$

9. Use the order of operations to solve the following problem:
$8 \times 2 - (4 - 2)^2 \div 2 =$ — 9. 14

10. Write an expression for the following statement:
Three less than five times a number. — 10. $5x - 3$

11. Evaluate the expression:
$4x + 5y - 3$ when $x = -2$ and $y = 4$ — 11. 9

12. Combine: $4x - 3x + 2x =$ — 12. $3x$

13. Use the distributive property to simplify the following expression:
$-4(x - 7) =$ — 13. $-4x + 28$

14. Solve the following equations.

a. $x - 12 = 16$ b. $-3x = -27$

c. $\frac{x}{13} = -2$ d. $6x - 14 = 4(x - 2)$

14a. $x = 28$
14b. $x = 9$
14c. $x = -26$
14d. $x = 3$

15. Two Little League baseball teams scored a total of 12 runs. The Seals scored three times as many runs as the Pups. How many runs did each team score? — 15. Pups = 3 runs; Seals = 9 runs

Glossary

A

Addends Numbers that are combined in the addition process.

Example: 8 + 9 = 17,

of which 8 and 9 are the addends.

Annually Yearly. Once a year.

Annual percentage rate (APR) True or effective annual interest rate charged by sellers. Required to be stated by Truth in Lending Act.

B

Balance The difference between totals.

Example:

(total revenues) – (total expenses).

Banker's Rule Time (exact time/360) in calculating simple interest.

Bank reconciliation Process of comparing the bank balance to the checkbook balance so adjustments can be made, checks outstanding, deposits in transit, and the like.

Bank statement Report sent by the bank to the owner of the checking account indicating checks processed, deposits made, and so on, along with beginning and ending balance.

Base Number that represents the whole 100%. It is the whole to which something is being compared. Usually follows word *of.*

Borrowing In subtraction, taking a unit of 10 from any number in the minuend and adding it to the next lower number in the minuend.

C

Cancellation Reducing process that is used to simplify the multiplication and division of fractions.

Example: $\frac{\cancel{4}^{1}}{8} \times \frac{1}{\cancel{4}_{1}}$

Cash discount Savings that result from early payment by taking advantage of discounts offered by the seller; discount is not taken on freight or taxes.

Chain discount Two or more trade discounts that are applied to the balance remaining after the previous discount is taken. Often called a *series discount.*

Checks Written documents signed by appropriate person that directs the bank to pay a specific amount of money to a particular person or company.

CM Abbreviation for *credit memorandum.* The bank is adding to your account. The CM is found on the bank statement.

Example: Bank collects a note for you.

Commissions Payments based on established performance criteria.

Common fraction A fraction expressed by two numbers, a numerator and a denominator.

Complement 100% less the stated percent.

Example: 18% ⇒ 82% is the complement (100% – 18%).

Compounding (FV) Calculating the interest periodically over the life of the loan and adding it to the principal.

Compound amount The total obtained at the end of the last compound period.

Compound interest The interest that is calculated periodically and then added to the principal. The next period the interest is calculated on the adjusted principal (old principal + interest).

Compound numbers The numbers 21 through 99. A quantity containing more than one unit.

Compound period The duration of time over which the compound interest is calculated.

Constants Numbers such as 3 or –7. Placed on right side of equation.

Conversion periods How often (a period of time) interest is calculated in the compounding process.

Example: Daily—each day; monthly—12 times a year; quarterly—every 3 months; semiannually—every 6 months.

D

Daily compounding Interest calculated on balance each day.

Decimal equivalent Decimal represents the same value of the fraction.

Example: $.05 = \frac{5}{100}$.

Decimal fraction Decimal representing a fraction. The denominator has a power of 10.

Decimal point Position located between units and tenths.

Decimals Numbers written with a decimal point.

Examples: 5.3, 18.22.

Deductions Amounts deducted from gross earnings to arrive at net pay.

Denominator The number of a fraction below the division line (bar).

Example: $\frac{8}{9}$, of which 9 is the denominator.

Difference The resulting answer from a subtraction problem.

Example: Minuend less subtrahend equals difference. $215 - 15 = 200$.

Digit Our decimal number system of 10 characters from 0 to 9.

Discount period Amount of time to take advantage of a cash discount.

Distributive Property Used to simplify a mathematical expression.

Example: $2(4x + 1) = 8x + 2$.

Dividend Number in the division process that is being divided by another.

Example: $5\overline{)15}$, in which 15 is the dividend.

Divisor Number in the division process that is being divided into another number.

Example: $5\overline{)15}$, in which 5 is the divisor.

DM Abbreviation for *debit memorandum.* The bank is charging your account. The DM is found on the bank statement.

E

Equation Math statement that shows equality for expressions or numbers or both.

Equivalent (fractional) Two or more fractions equivalent in value.

Exponent The number that indicates how many times the base occurs as a factor.

Expression Any combination of numbers and variables linked by arithmetic operations.

F

Factors Numbers multiplied together.

Example: $2 \times 3 = 6$. Two and three are called *factors.*

Federal Insurance Contribution Act (FICA) Percent on base amount of each employee's salary. FICA taxes used to fund retirement, disability income, Medicare, and so on.

Federal withholding tax *See* income tax.

Fraction Expresses a part of a whole number.

Example: $\frac{5}{6}$ expresses 5 parts out of 6.

G

Greatest common divisor The largest possible number that will divide evenly into both the numerator and denominator.

Gross pay Wages before deductions.

Grouping symbols Parentheses ( ) are used as grouping symbols. The expression within the set of parentheses is treated as a single number.

H

Higher terms Expressing a fraction with a new numerator and denominator that is equivalent to the original.

Example: $\frac{2}{9} \Rightarrow \frac{6}{27}$.

I

Improper fraction Type of fraction when numerator is equal to or greater than the denominator.

Examples: $\frac{6}{6}, \frac{14}{9}$.

Income tax (FIT) Tax depends on allowances claimed, marital status, and wages earned.

Interest Principal × rate × time.

Inverting Interchanging the numerator and the denominator of a fraction to finds its reciprocal.

Example: $\frac{2}{3}$ gives $\frac{3}{2}$.

Invoice Document recording purchase and sales transactions.

L

Least common denominator (LCD) Smallest whole number into which denominator of two or more fractions will divide evenly.

Example: $\frac{2}{3}$ and $\frac{1}{4}$ LCD = 12.

Like fractions Proper fractions with the same denominators.

Like terms Terms that are made up with the same variable.

Example: $a + 2a + 3a = 6a$.

List price Suggested retail price paid by customers.

Long division The division process where the multiplication and subtraction steps are written.

Lowest terms Expressing a fraction when no number divides evenly into the numerator and denominator except the number 1.

Example: $\frac{5}{10} \Rightarrow \frac{1}{2}$.

M

Minuend In a subtraction problem, the larger number from which another is subtracted.

Example: $50 - 40 = 10$.

Mixed decimal Combination of a whole number and a decimal.

Examples: 2.14, 17.39

Mixed numbers Numbers written as a whole number and a proper fraction.

Examples: $2\frac{1}{4}, 3\frac{8}{9}$.

Monthly One each month. Some employers pay employees monthly.

Multiple The product of a given number and another number.

Example: In $2 \times 7 = 14$, fourteen is the multiple.

Multiplicand The first (or top) number being multiplied in a multiplication problem.

Example:

Product = Multiplicand × Multiplier
40 = 20 × 2.

Multiplier The second (or bottom) number doing the multiplication in a problem.

Product = Multiplicand × Multiplier
40 = 20 × 2.

N

Negative numbers Numbers less than zero. A collection of numbers with negative signs.

Examples: $-5, -\frac{1}{2}$.

Net The value obtained after deducting all expenses or deductions.

Net pay *See* net wages.

Net price List price less amount of trade discount. The net price is before any cash discount.

Net wages Gross pay less deductions.

Numeral A symbol used for a number.

Numerator Number of a fraction above the division line (bar).

Example: $\frac{8}{9}$, in which 8 is the numerator.

O

Order of operations A specific order in which arithmetic operations must be carried out when more than one operation is to be done. The order is: parentheses, exponents, multiplication or division, addition or subtraction.

Outstanding checks Checks written but not yet processed by the bank before bank statement preparation.

P

Partial products Numbers between multiplier and product.

Percent Stands for hundredths.

Example: 4% is 4 parts of a hundred, or $\frac{4}{100}$.

Percent decrease Calculated by decrease in price over original amount.

Percent increase Calculated by increase in price over original amount.

Periods Number of years multiplied by the number of times compounded per year (*see* conversion period).

Place value The digit value that results from its position in a number.

Portion Amount, part, or portion that results from multiplying the base by the rate. Not expressed as a percent, it is expressed as a number.

Positive numbers Numbers greater than zero. Collection of numbers with a positive sign.

Examples: $+\frac{5}{7}, +9$.

Prime numbers Numbers (larger than 1) that are only divisible by itself and 1.

Examples: 2, 3, 5.

Principal Amount of money that is originally borrowed, loaned, or deposited.

Product Answer of a multiplication process.

Example:

Product = Multiplicand × Multiplier
50 = 5 × 10.

Proper fractions Fractions with numerator less than denominator.

Example: $\frac{5}{9}$.

Q

Quarterly Occurring at intervals of 3 months.

Quotient The answer of a division problem.

R

Rate Percent that is multiplied by the base that indicates what part of the base to compare to. Rate is not a whole number.

Reciprocal of a fraction The interchanging of the numerator and the denominator.

Example: $\frac{6}{7} \Rightarrow \frac{7}{6}$.

Reduce To change the form of a fraction without changing its value.

Remainder Leftover amount in division.

Repeating decimals Decimal numbers that repeat themselves continuously and therefore do not end.

Root A quantity that, when multiplied by itself a specified number of times, generates a given quantity.

Rounding decimals Reducing the number of decimals to an indicated position.

Example: 59.59 ⇒ 59.6 to nearest tenth.

Rounding whole numbers all the way Process to estimate actual answer. When rounding all the way, only one nonzero digit is left. Rounding all the way gives the least degree of accuracy.

Examples: 1,251 to 1,000; 2,995 to 3,000.

S

Selling price Cost plus markup equals selling price.

Semiannually Twice a year.

Simple interest Interest is only calculated on the principal. In $I = P \times R \times T$, the interest plus original principal equals the maturity value of an interest-bearing note.

Simple interest formula

$$\text{Interest} = \text{Principal} \times \text{Rate} \times \text{Time}$$

$$\text{Principal} = \frac{\text{Interest}}{\text{Rate} \times \text{Time}}$$

$$\text{Rate} = \frac{\text{Interest}}{\text{Principal} \times \text{Time}}$$

$$\text{Time} = \frac{\text{Interest}}{\text{Principal} \times \text{Rate}}$$

Short division The division process where the multiplication and subtraction steps are not written.

Similar terms *See* like terms.

Stocks Ownership shares in the company sold to buyers who receive stock certificates.

Straight commission Wages are based as a percent of the value of goods sold.

Subtrahend Smaller number in a subtraction problem that is being subtracted from another.

Example: 150 – 30 = 120.

Sum Total in the adding process.

T

Time Expressed as years or fractional years. Used to calculate the simple interest.

Trade discount amount List price less net price.

U

Unlike fraction Proper fractions with different denominators.

V

Variable commission scale Company pays different commission rates for different levels of net sales.

Variables Letters or symbols that represent unknowns.

W

Whole number Numbers which don't contain a decimal or fraction part.

Examples: 10, 55, 92.

Withholding Amount or deduction from one's paycheck.

APPENDIX

Answers to Selected Problems

CHAPTER 1

Unit 1

Practice 1: seven hundred eighty-three thousand, six hundred ninety-six

Practice 2: 3,195,202

1. thirty-seven
4. one thousand seventy-three
8. two million, three hundred thousand, five hundred eighty-one
12. sixteen billion, one hundred forty-seven million, three hundred seventeen thousand, two hundred ninety-six
16. 62,000,000
20. 39,435,000,000
24. 510,846,182

Business Applications:

27. seventeen million, four hundred twenty thousand
28. forty-six million, six hundred twenty thousand, three hundred twenty-seven
29. $14,732
30. $3,643

Unit 2

Practice 1: 7,500

Practice 2: 41,000

Practice 3: 16,000

1. 90
4. 63,000
8. 28,500
12. 4,700,000
16. 14,000
20. 520,000

Business Applications:

24. $750,000
25. 3,200 shares

Unit 3

Practice 1: 9

Practice 2: 12

1. 8, 13, 11, 17, 11, 2, 14
4. 7, 12, 12, 10, 3, 14, 9
8. 16, 20, 23, 17, 16, 18, 21

Business Applications:

9. 18 hours
10. $14

Unit 4

Practice 1: 1,304

Practice 2: 628

1. 1,178
4. 1,065
8. 923
12. 22,656
16. 2,045
20. 2,266
24. 2,600,830
28. 1,286
32. 832

Business Applications:

34. 486 patrons
35. 109 cases

Unit 5

Practice 1: 360,000

Practice 2: 40,300

1. 1,500
4. 2,409,000
8. 18,130

Business Applications:

9. $1,110
10. 7,300,000

Unit 6

Practice 1: 423

Practice 2: 215

1. 55
4. 22
8. 112
12. 2,220
16. 21,104
20. 82,513
24. 52
28. 603
32. 65,222
36. 369,574

Business Applications:

39. $61,399
40. 1,013 textbooks

Unit 7

Practice 1: 275

Practice 2: 5,778

Practice 3: 3,237

1. 252
4. 235
8. 22
12. 67
16. 3,316
20. 34,740

Business Applications:

23. $14
24. $134

Unit 8

Practice 1: 62,116 square miles

Practice 2: $413.00

1. 22,791 square feet
3. 8,718 people
5. 5,281 people
7. 9 bags
9. 791 pounds
11. 1,494 calls
13. $375.00
15. 6,448 people
17. $1,673.00
19. soccer; 500 children

Unit 9

×	1	2	3	4	5	6	7	8	9	10	11	12
1	1	2	3	4	5	6	7	8	9	10	11	12
2	2	4	6	8	10	12	14	16	18	20	22	24
3	3	6	9	12	15	18	21	24	27	30	33	36
4	4	8	12	16	20	24	28	32	36	40	44	48
5	5	10	15	20	25	30	35	40	45	50	55	60
6	6	12	18	24	30	36	42	48	54	60	66	72
7	7	14	21	28	35	42	49	56	63	70	77	84
8	8	16	24	32	40	48	56	64	72	80	88	96
9	9	18	27	36	45	54	63	72	81	90	99	108
10	10	20	30	40	50	60	70	80	90	100	110	120
11	11	22	33	44	55	66	77	88	99	110	121	132
12	12	24	36	48	60	72	84	96	108	120	132	144

×	7	4	1	3	11	9	5	12	8	2	6	10
8	56	32	8	24	88	72	40	96	64	16	48	80
4	28	16	4	12	44	36	20	48	32	8	24	40
11	77	44	11	33	121	99	55	132	88	22	66	110
6	42	24	6	18	66	54	30	72	48	12	36	60
9	63	36	9	27	99	81	45	108	72	18	54	90
12	84	48	12	36	132	108	60	144	96	24	72	120
1	7	4	1	3	11	9	5	12	8	2	6	10
3	21	12	3	9	33	27	15	36	24	6	18	30
10	70	40	10	30	110	90	50	120	80	20	60	100
2	14	8	2	6	22	18	10	24	16	4	12	20
7	49	28	7	21	77	63	35	84	56	14	42	70
5	35	20	5	15	55	45	25	60	40	10	30	50

Unit 10

Practice 1: 78

Practice 2: 16,748

1. 304
4. 81
8. 3,654
12. 4,165
16. 24,276
20. 32,936
24. 93,708

Business Applications:

26. $3,900
27. 378 miles

Unit 11

Practice 1: 2,445,210

Practice 2: 1,331,200,000

Practice 3: 890
8,900
89,000
1,780
17,800
178,000
2,670
26,700
267,000

1. 313,992
4. 336,630
8. 15,062,063
12. 408,200
16. 2,226,000
20. 34,263,000,000
24. 97,290,000
28. 297,500
32. 1,487,500
36. 47,256,000

Business Applications:

37. $2,100
38. 45,000 balls

Unit 12

1. 3; 1; 9; 4; 2; 5; 6; 8
4. 9; 5; 3; 4; 4; 5; 6; 6
8. 1; 5; 9; 7; 1; 3; 9; 8

Business Applications:

9. 9 parts
10. 7 hours

Unit 13

Practice 1: 823

Practice 2: 1,509

1. 2,431
4. 1,111
8. 8,111
12. 8,105
16. 91,205
20. 9,310 R2
24. 1,784 R3
28. 9,003 R6

Business Applications:

29. 252 gallons
30. $32,500

Unit 14

Practice 1: 5,628
Practice 2: 2,204 R24

1. 11,154
4. 76,419
8. 452 R3
12. 33,369 R13
16. 4,780 R7
20. 78,540
24. 912

Business Applications:

28. $500
29. $38,500

Unit 15

Practice 1: 365
Practice 2: 38 R516

1. 650
4. 375
8. 320 R762
12. 90,213 R640
16. 238 R13

Business Applications:

19. 600 payments
20. $8,900

Unit 16

Practice 1: 24 decks
Practice 2: 4 minutes

1. $1,120
3. $576
5. 503,308 people
7. 47 appointments
9. $624
11. $525
13. 26 miles per gallon
15. $1,314,600

Chapter 1 Test

1. six million, seven hundred thirteen thousand, four hundred seventy-six
2. 261,392
3. 48,000
4. 497
5. 519
6. 18,000
7. 2,701 tickets
8. 42,244

9a. 33,264,000
9b. 332,640,000
9c. 3,326,400,000

10. 1,375
11. 162 R273
12. $16,375

CHAPTER 2

Unit 17

Practice 1: Proper: $\frac{3}{4}$, $\frac{14}{16}$, $\frac{6}{7}$
Improper: $\frac{6}{5}$, $\frac{81}{3}$, $\frac{8}{8}$
Mixed: $1\frac{4}{9}$, $3\frac{9}{12}$

Practice 2: $\frac{6}{6}$, $\frac{49}{49}$, $\frac{166}{166}$

Practice 3: $\frac{19}{20}$, $8\frac{3}{17}$, $\frac{6}{4}$

1. P
4. P
8. I
12. P
16. 1
20. P
24. $2\frac{2}{7}$
28. $\frac{15}{15}$

Unit 18

Practice 1: $5\frac{3}{4}$, 9
Practice 2: $\frac{31}{4}$

1. $2\frac{6}{9}$
4. $13\frac{1}{4}$
8. $8\frac{3}{10}$
12. $7\frac{5}{6}$
16. $11\frac{2}{3}$
20. $11\frac{3}{4}$
24. $\frac{20}{17}$
28. $\frac{45}{7}$
32. $\frac{27}{4}$
36. $\frac{288}{20}$
40. $\frac{77}{6}$

Unit 19

Practice 1: $\frac{1}{2}$
Practice 2: $4\frac{4}{9}$

1. $\frac{5}{6}$
4. $\frac{1}{4}$
8. $\frac{1}{4}$
12. $\frac{5}{7}$
16. $1\frac{7}{9}$
20. $3\frac{1}{4}$
24. $\frac{2}{3}$
28. $18\frac{8}{9}$
32. $10\frac{7}{12}$

Business Applications:

34. $\frac{1}{8}$
35. $\frac{3}{35}$

Unit 20

Practice 1: a. 16
b. 60

1. 25
4. 27
8. 54
12. 400
16. 90
20. 160
24. 48
28. 49

Business Applications:

29. 30 monitors
30. $\frac{2,520}{3,500}$

Unit 21

Practice 1: a. $\frac{9}{11}$
b. $8\frac{3}{7}$
c. $2\frac{1}{8}$

1. $\frac{4}{7}$
4. $\frac{3}{4}$

8. $1\frac{1}{7}$
12. $13\frac{3}{5}$
16. $8\frac{1}{3}$
20. $14\frac{1}{5}$

Business Applications:

24. $24\frac{1}{2}$
25. 62

Unit 22

Practice 1: a. $2 \times 2 \times 2 \times 3 \times 3$
b. $2 \times 2 \times 2 \times 3 \times 3 \times 5$

Practice 2: 120

1. $2 \times 2 \times 2 \times 2 \times 2 \times 2$
4. $3 \times 3 \times 5$
8. $2 \times 2 \times 2 \times 2 \times 2 \times 3$
12. 24
16. 192
20. 480

Unit 23

Practice 1: $10\frac{1}{30}$

1. $\frac{3}{4}$
4. $1\frac{1}{4}$
8. $2\frac{5}{24}$
12. $21\frac{1}{12}$
16. $30\frac{2}{3}$
20. $14\frac{13}{24}$

Business Applications:

21. $23\frac{5}{8}$
22. $\frac{23}{40}$ of the shipment

Unit 24

Practice 1: $3\frac{1}{3}$

1. $\frac{3}{5}$
4. $\frac{1}{3}$
8. $\frac{2}{3}$
12. $\frac{4}{15}$
16. $9\frac{1}{4}$
20. $6\frac{3}{4}$
24. $8\frac{1}{12}$
28. $4\frac{3}{10}$

Business Applications:

29. $1\frac{1}{2}$ hours
30. $3\frac{1}{4}$ yards

Unit 25

Practice 1: $\frac{23}{77}$

Practice 2: $2\frac{8}{15}$

Practice 3: $10\frac{1}{3}$

1. $\frac{1}{4}$
4. $\frac{7}{12}$
8. $\frac{1}{6}$
12. $1\frac{1}{4}$
16. $1\frac{5}{8}$
20. $7\frac{1}{15}$
24. $1\frac{7}{8}$

Business Applications:

26. $5\frac{7}{8}$
27. $2\frac{1}{4}$ hours

Unit 26

Practice 1: $\frac{17}{24}$ of the book

Practice 2: $4\frac{3}{8}$ acres

1. $2\frac{3}{4}$ pounds
3. $4\frac{1}{16}$ inches
5. $6\frac{1}{24}$ hours
7. $6\frac{1}{8}$ miles
9. $20\frac{23}{24}$ gallons
11. $20\frac{5}{6}$ gallons
13. $434\frac{3}{8}$ square feet
15. $495\frac{4}{5}$ feet
17. $1\frac{17}{30}$ hours
19. $\$1\frac{7}{8}$ per share

Unit 27

Practice 1: $\frac{1}{18}$

Practice 2: $\frac{8}{99}$

1. $\frac{7}{12}$
4. $\frac{4}{15}$
8. $\frac{2}{5}$
12. $\frac{7}{10}$
16. $\frac{16}{35}$
20. $\frac{2}{9}$
24. $\frac{1}{140}$
28. $\frac{2}{49}$

Business Applications:

29. $\frac{1}{2}$ of the sales force
30. $\frac{3}{16}$ of a ton

Unit 28

Practice 1: 45

Practice 2: $3\frac{6}{19}$

1. $\frac{15}{16}$
4. 3
8. $38\frac{6}{7}$
12. $12\frac{4}{5}$
16. $1\frac{41}{44}$
20. 232
24. $228\frac{2}{3}$

Business Applications:

28. 234 auto claims
29. $4\frac{1}{3}$ acres

Unit 29

Practice 1: $\frac{1}{13}$

Practice 2: $\frac{7}{9}$

1. $\frac{9}{4}$
4. 13
8. $2\frac{1}{2}$
12. $\frac{4}{5}$
16. 1
20. $1\frac{1}{2}$
24. $\frac{9}{11}$

Business Applications:

25. 24 screws
26. 14 sections

Unit 30

Practice 1: $\frac{1}{9}$

Practice 2: $1\frac{9}{10}$

1. $\frac{7}{22}$
4. $1\frac{6}{55}$
8. $1\frac{17}{19}$
12. 56
16. $\frac{7}{8}$
20. $\frac{9}{40}$
24. $\frac{1}{2}$

Business Applications:

25. 400 lots
26. 4,960 bags

Unit 31

Practice 1: 27 girls

Practice 2: 150 sneakers

Practice 3: 49 packages

1. $\frac{1}{2}$ inch
3. $13\frac{1}{8}$ pounds
5. 10 truckloads
7. $\frac{3}{8}$ pound
9. 12 females
11. $48\frac{3}{10}$ miles per hour
13. 16 runs
15. $11\frac{1}{2}$ miles
17. 22 slices
19. 42 tiles

Chapter 2 Test

1. a. I
   b. 1
   c. I
   d. P
   e. M
2. $6\frac{4}{7}$
3. $\frac{33}{5}$
4. a. $\frac{1}{3}$
   b. $3\frac{1}{4}$
5. $\frac{12}{52}$
6. 2
7. $9\frac{19}{48}$
8. $\frac{13}{24}$
9. $6\frac{9}{20}$
10. $4\frac{1}{8}$ tons
11. $1\frac{11}{12}$
12. $1\frac{1}{3}$
13. 92 lots

CHAPTER 3

Unit 32

Practice 1: sixteen ten thousandths

forty-one and seventeen thousandths

Practice 2: .0081

50.93

1. .3
4. 17.089
8. .0957
12. 20.0002
16. four hundred seven thousandths
20. sixty-nine ten thousandths
24. ten and seven ten thousandths

Business Applications:

27. 2.007 acres
28. forty-one and two tenths hours

Unit 33

Practice 1: .17

.84

5.61

14.29

Practice 2: 18.0

$33.77

120.00

1. 4.4
4. 42.9
8. .3
12. .362
16. .24
20. 10.08
24. $5.00
28. $965.08

Business Applications:

29. 506.8 gallons
30. 105 yards

Unit 34

Practice 1: 253.6819

Practice 2: 344.7826

1. 33.993
4. 1,712.156
8. 3,273.2
12. 124.500763
16. 5.203
20. 18.61
24. 450.713
28. .049235

Business Applications:

29. 549.965 pounds
30. $3,996.20

Unit 35

Practice 1: $1,161.60

Practice 2: $2,873.65

1. 841.68 square feet
3. $194.80
5. 1,167.22 square feet
7. 17.128 gallons
9. 4.347 centimeters
11. 505.7486 acres
13. 2.1875 feet
15. $377.40
17. 19.825 miles
19. 38.285 centimeters

Unit 36

Practice 1: 11.804

Practice 2: .0016269

1. 241.66
4. .73107
8. .031812
12. 305.3798
16. 6,076.2
20. .009022
24. 134.55442

Business Applications:

25. $247.76
26. $421,875.00

Unit 37

Practice 1: .732
24.31
Practice 2: 42.16
Practice 3: 30.92

1. .172
4. .03
8. .045
12. 3.5
16. 470
20. 5.30
24. 2.946
28. .3202

Business Applications:

29. 625.155 pounds
30. $15

Unit 38

Practice 1: 132.8
1,328
13,280
Practice 2: 8.91
.891
.0891

1. 73.8
4. 630
8. 3,247
12. .71
16. 696
20. 2,710
24. .00074
28. 7.38
32. 003896
36. .00119
40. .01399

Unit 39

Practice 1: $41.93
Practice 2: 21 miles per gallon

1. 263.8 inches
3. 12 blocks
5. 19 months
7. $.17
9. 705 people
11. $1,779.73
13. 3.175
15. 8 hours
17. 5,625 milligrams
19. $272.20

Unit 40

Practice 1: .07
.014
.0072
.072
Practice 2: .08
.111
Practice 3: 16.25

1. .16
4. .0014
8. 6.5
12. .25
16. 4.2
20. 11.375
24. .88
28. .2
32. .14
36. 8.4

Business Applications:

39. 20.75
40. +.375

Unit 41

Practice 1: $\frac{11}{50}$
$\frac{7}{1,000}$
$33\frac{4}{25}$
Practice 2: $\frac{13}{80}$

1. $\frac{9}{100}$
4. $\frac{1}{25}$
8. $\frac{13}{100}$
12. $4\frac{8}{25}$
16. $4\frac{157}{500}$
20. $\frac{481}{100,000}$
24. $68\frac{3}{16}$
28. $5\frac{5}{8}$
32. $\frac{41}{5,000}$

Business Applications:

34. $6\frac{3}{4}$ points
35. $3\frac{3}{5}$ hours

Unit 42

Practice 1: $329.56
Practice 2: $5,752.50

1. $354.30
3. $235.69
5. $10,426.80
7. $433.88
9. $8,677.20
11. $1,575.16
13. $21.45
15. $1,786.48

Unit 43

Practice 1: $289.25
Practice 2:

Gross Wages	$894.00
Federal Withholding Tax	$118.00
Social Security Tax	$55.43
Medicare Tax	$12.96
Total Deductions	$186.39
Net Pay	$707.61

Employee Name	Hours Worked S	M	T	W	Th	F	S	Total Hours	Hourly Rate	Gross Wage
Biller, Norman	0	8	8	8	7	3	4	38	$22.50	$855.00
Carter, Susan	0	8	8	8	7.5	5	4	40.5	$21.35	$864.68
Chidner, Jill	0	8	7.5	8	9	6	4	42.5	$22.50	$956.25
Dobbins, Edward	0	8	8	8	8	8	6.5	46.5	$23.75	$1,104.38
Dimitri, Sal	0	8	8	8	7.5	8	4	43.5	$22.50	$978.75
Fortin, Nelson	0	8	8	6	0	9.5	4	35.5	$21.35	$757.93
Flashberg, Amy	0	8	8	8	8	8	5.5	45.5	$21.35	$971.43
Harner, Stephen	4	7.5	8	8	8	0	0	35.5	$23.75	$843.13
Husson, Valerie	0	8	8	8	8	8.5	4.5	45	$22.50	$1.012.50
Koney, Stuart	5	8	8	7.5	8	9.5	0	46	$22.50	$1,035.00

Name	Withholding Allowance and Marital Status	Hours Worked	Hourly Rate	Gross Wages	Deductions				Net Pay
					Federal Withholding Tax	Social Security Tax	Medicare Tax	Total Deductions	
Atwell	S–0	37.5	$19.25	$721.88	$137.00	$44.76	$10.47	$192.23	$529.65
Autens	M–1	35.5	$21.25	$754.38	$89.00	$46.77	$10.94	$146.71	$607.67
Ayers	M–2	38	$21.25	$807.50	$89.00	$50.07	$11.71	$150.78	$656.72
Brown	S–0	39	$19.50	$760.50	$148.00	$47.15	$11.03	$206.18	$554.32
Buckley	S–0	40	$18.75	$750.00	$145.00	$46.50	$10.88	$202.38	$547.62
Green	S–1	37.5	$20.50	$768.75	$136.00	$47.66	$11.15	$194.81	$573.94
Harkins	M–2	40	$20.50	$820.00	$92.00	$50.84	$11.89	$154.73	$665.27
Hill	M–2	40	$21.25	$850.00	$97.00	$52.70	$12.33	$162.03	$687.97
Johnson	M–3	40	$21.25	$850.00	$90.00	$52.70	$12.33	$155.03	$694.97
Miller	M–1	38.5	$20.50	$789.25	$93.00	$48.93	$11.44	$153.37	$635.88

Chapter 3 Test

1. 253.0027
2. Twenty-nine and seventy-one hundredths
3. 3.1
4. 171.5636
5. 10.537
6. $207.59
7. 134.216892
8. 60.105
9a. 42.3
9b. 423
9c. 4,230
10a. 1.937
10b. .1937
10c. .01937
11. $188.30
12. 16.875
13. $4\frac{4}{25}$

CHAPTER 4

Unit 44

Practice 1: 41%
4.3%
1,140%

Practice 2: 25%
125%
83.3%

1. 4%
4. 289%
8. 70%
12. 109%
16. 2,200%
20. 375%
24. 175%
28. 314.3%
32. 83.3%

Business Applications:

36. 20%
37. 30%

Unit 45

Practice 1: .72
.0615
.002
.022

Practice 2: $\frac{7}{50}$
$\frac{37}{1,000}$
$\frac{9}{400}$

1. .11
4. 1.14
8. .02125
12. .0275
16. .0108
20. .166
24. $1\frac{21}{50}$
28. $\frac{3}{2,500}$
32. $\frac{9}{25}$
36. $\frac{39}{800}$

Business Applications:

39. .05
40. $\frac{17}{250}$

Unit 46

Practice 1:

Base = 200, Rate = 22%, Portion = 44

Base = 32, Rate = 25%, Portion = 8

Base = 36, Rate = unknown, Portion = 12

Base = 200, Rate = 13%, Portion = unknown

Practice 2: 10
4%
420

Practice 3: 77
40
200%

1. b = 100, r = 22%, p = 22
4. b = 80, r = 12%, p = 9.6
8. b = 50, r = 16.2%, p = 8.1
12. 20.8
16. 66.7%
20. 2.56
24. 320
28. 10.5
32. 250

Business Applications:

34. 46,872 passengers
35. 10.4%

Unit 47

Practice 1: 30 pieces
Practice 2: 33%
Practice 3: 450 students

1. 64.8 pounds
3. $1,275
5. 64%
7. 5%
9. $10,000
11. 75 Boy Scouts
13. 22.2%
15. 10 windows
17. $1,050.00
19. 125 members

Unit 48

Practice 1: $60.22
Practice 2: 6.3%
Practice 3: 49 applications
Practice 4: 150 games

1. $19.52
3. 5%
5. 38 houses
7. 9,660 tourists
9. 11,000 seats
11. 19%
13. $770.40
15. 500 students
17. 30.96 miles per hour
19. $17,321.48

Unit 49

Practice 1: $7,295
Practice 2: $227.50
Practice 3: 6%

1. $13,170
3. $1,351.44

5. 9.5%
7. $1,472.00
9. $5,171.25
11. $12,000.00
13. $800.00
15. $2,316.00
17. $30,220.00
19. 5%

Unit 50

Practice 1:

Trade Discount = $336;
Net Price = $2,464

Practice 2:

Net Price Rate = .72675
Net Price = $10,901.25
Trade Discount = $4,098.75

Practice 3: $525.00

3. $936.00
5. 25%
7. $160.71
9. $1,206.58
11. Trade Discount = $2,583.00
    Net Price =$4,617.00

1.

List Price	Trade Discount Price	Trade Discount	Net Price
$1,250	40%	$500.00	$750.00
$1,500	30%	$450.00	$1,050.00
$6,005	45%	$2,702.25	$3,302.75
$1,850	15%	$277.50	$1,572.50
$5,775	35%	$2,021.25	$3,753.75
$680	22%	$149.60	$530.40
$2,780	16.5%	$458.70	$2,321.30
$3,750	7.5%	$281.25	$3,468.75
$4,900	$13\frac{1}{4}$%	$649.25	$4,250.75
$7,450	$18\frac{1}{2}$%	$1,378.25	$6,071.75

2.

List Price	Chain Discount	Net Price Rate	Net Price	Trade Discount
$4,500	20/10	.72	$3,240.00	$1,260.00
$3,250	25/10	.675	$2,193.75	$1,056.25
$5,775	30/20	.56	$3,234.00	$2,541.00
$1,865	25/10	.675	$1,258.88	$606.12
$18,000	10/5/2	.8379	$15,082.20	$2,917.80
$25,000	20/15/10	.612	$15,300.00	$9,700.00
$12,000	25/20/5	.57	$6,840.00	$5,160.00
$15,000	10/8/3	.80316	$12,047.40	$2,952.60
$105,000	20/15/12	.5984	$62,832.00	$42,168.00
$32,000	20/15/10	.612	$19,584.00	$12,416.00

Unit 51

Practice 1:

Cash Discount = $29.20

Net Amount Paid = $1,430.80

Practice 2:

Cash Discount = $40.00

Net Amount Paid = $1,960.00

1. $5,820.00
3. $5,250.00
5. $8,245.00

Amount of Invoice	Terms	Cash Discount	Net Amount Paid
$1,995	3/10, n/30	$59.85	$1,935.15
$2,025	2/15, n/30	$40.50	$1,984.50
$3,199	4/10, n/30	$127.96	$3,071.04
$4,220	2/10, n/30	$84.40	$4,135.60
$1,650.45	5/10, n/30	$82.52	$1,567.93
$1,735.90	2/15, n/45	$34.72	$1,701.18
$5,746	2/10, n/30	$114.92	$5,631.08
$916	3/10, n/45	$27.48	$888.52
$2,488	4/10, n/30	$99.52	$2,388.48
$649	5/10, n/30	$32.45	$616.55

Date of Invoice	Amount of Invoice	Terms	Date Paid	Cash Discount	Net Amount Paid
April 13	$4,800	3/10, n/30	April 18	$144.00	$4,656.00
February 3	$1,600	5/10, n/30	February 28	0	$1,600.00
June 6	$2,946	2/10, 1/15, n/30	June 18	$29.46	$2,916.54
October 12	$626	3/10, 2/15, n/30	October 25	$12.52	$613.48
March 26	$1,980	1/10, n/30	April 4	$19.80	$1,960.20
October 10	$4,770	2/10, 1/30, n/60	October 28	$47.70	$4,722.30
January 15	$2,015	3/5, 2/20, n/60	February 2	$40.30	$1,974.70
November 3	$1,675	2/10, 1/15, n/30	November 17	$16.75	$1,658.25
August 17	$1,222	3/10, 2/15, n/30	August 31	$24.44	$1,197.56
September 12	$850	2/10, 1/15, n/30	September 19	$17.00	$833.00
July 27	$4,500	2/10, 1/15, n/30	August 6	$90.00	$4,410.00
November 24	$6,750	3/10, 2/15, n/30	December 5	$135.00	$6,615.00
June 13	$7,800	2/10, 1/30, n/60	July 12	$78.00	$7,722.00
August 21	$844	5/50, 2/30, n/60	September 17	$16.88	$827.12
April 6	$5,200	2/10, 1/30, n/60	May 26	0	$5,200.00

Unit 52

Practice 1: Interest = \$18.75
Principal = \$1,600
Rate = 8%
Time = 1.5 years

Practice 2: \$68.25

Interest	Principal	Rate	Time
\$7.25	\$1,450	3%	60 days
\$480	\$4,000	6%	2 years
\$227.50	\$3,500	$6\frac{1}{2}$%	1 year
\$1,350	\$3,750	12%	3 years
\$98.80	\$760	$3\frac{1}{4}$%	4 years
\$20.00	\$1,000	6%	120 days
\$800	\$2,500	8%	4 years
\$9.00	\$800	$4\frac{1}{2}$%	90 days
\$1,350	\$3,750	12%	3 years
\$26.25	\$125	7%	3 years
\$41.25	\$250	5.5%	3 years
\$148.50	\$2,140.54	$9\frac{1}{4}$%	9 months
\$100.00	\$4,000	5%	6 months
\$495	\$1,500	$5\frac{1}{2}$%	6 years
\$6	\$240	5%	6 months

1. \$3,135.00
3. \$8,400.00
5. \$2,420.00
7. \$6,087.00
9. \$1,680.00

Unit 53

Practice 1:

Compound Interest = $1,511.75

Compound Amount = $4,011.75

Practice 2:

Compound Interest = $3,098.40

Compound Amount = $6,098.40

Time	Principal	Rate	Compounded	Interest Rate per Compounding Period	# of Periods	Table Factor	Compound Amount	Compound Interest
3 yrs.	$2,000	8%	semiannually	4%	6	1.2653	$2,530.60	$530.60
2 yrs.	$1,500	4%	quarterly	1%	8	1.0829	$1,624.35	$124.35
4 yrs.	$2,500	6%	annually	6%	4	1.2625	$3,156.25	$656.25
5 yrs.	$800	6%	quarterly	1.5%	20	1.3469	$1,077.52	$277.52
6 yrs.	$5,000	8%	semiannually	4%	12	1.6010	$8,005.00	$3,005.00
8 yrs.	$3,000	10%	semiannually	5%	16	2.1829	$6,548.70	$3,548.70
10 yrs.	$2,500	7%	annually	7%	10	1.9672	$4,918.00	$2,418.00
6 yrs.	$1,800	12%	quarterly	3%	24	2.0328	$3,659.04	$1,859.04
4 yrs.	$4,000	8%	quarterly	2%	16	1.3728	$5,491.20	$1,491.20
7 yrs.	$6,000	6%	quarterly	1.5%	28	1.5172	$9,103.20	$3,103.20
3 yrs.	$2,000	12%	semiannually	6%	6	1.4185	$2,837.00	$837.00
10 yrs.	$15,000	14%	semiannually	7%	20	3.8697	$58,045.50	$43,045.50
12 yrs.	$9,000	8%	annually	8%	12	2.5182	$22,663.80	$13,663.80
10 yrs.	$2,000	10%	semiannually	5%	20	2.6533	$5,306.60	$3,306.60
7 yrs.	$3,500	12%	quarterly	3%	28	2.2879	$8,007.65	$4,507.65

1. $7,600.80
3. $33,636.51
5. $8,385.50
7. yes; $48,867.00
9. $14,882.26

Chapter 4 Test

1a. 53%
1b. .6%
1c. 12.5%
1d. 77.8%
2a. .18
2b. .0414
3a. $\frac{37}{100}$
3b. $\frac{37}{150}$
4. Base = 250; Portion = 65; Rate = 26%
5. 40%
6. 20
7. 59.4
8. $220,000.00
9. 33.3%
10. $825.00
11. $936.00
12. $40.00
13. $18,000.00
14a. 4 periods
14b. 2%
14c. $16.48

CHAPTER 5

Unit 54

Practice 1:

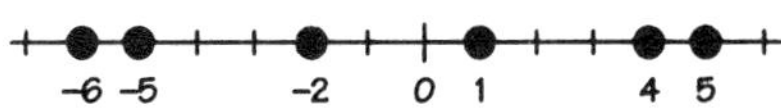

Practice 2: negative
negative
positive
positive

Practice 3: 14, $-\frac{7}{8}$, 5.2

1.

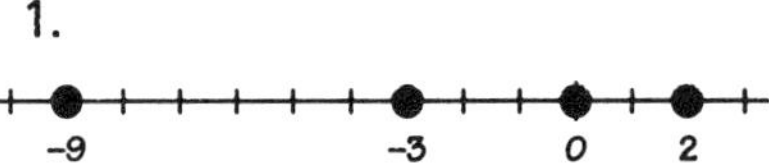

4.

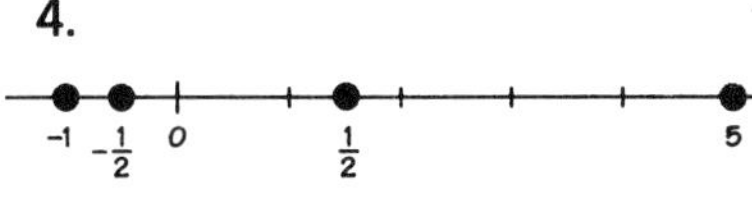

8. negative
12. negative
16. 32
20. 22.2
24. –30

Unit 55

Practice 1: –26
Practice 2: –8
Practice 3: –50
Practice 4: 41
1. –10
4. –45
8. –35
12. 5
16. 32
20. –42
24. 6
28. 60
32. 17
36. 22
40. –24

Business Applications:
41. –$34
42. $34
46. 16
50. 52

Unit 56

Practice 1: 45, –42
Practice 2: 2, –4, 1
Practice 3: –480, $-\frac{1}{21}$
1. –42
4. –45
8. 63
12. –10
16. –432
20. $\frac{1}{8}$
24. –6
28. –1
32. –4
36. 2
40. 1
44. $\frac{1}{7}$
48. $-\frac{3}{49}$

Unit 57

Practice 1: 64, 81, 10, 6
Practice 2: 45
1. 9
4. 4
8. 1,000
12. 6
16. 169
20. 8
24. 24
28. 36
32. –7

Unit 58

Practice 1:
$12 + x$; $4x$; $(x + 3) - 1$
Practice 2: 42
1. $x + 7$
4. $6x + 2$
8. $2x - 2$
12. 6
16. 22
20. 19

Business Applications:
24. $2x$
25. $4x$

Unit 59

Practice 1: $x = 12$
Practice 2: $x = 23$
Practice 3: $x = 10$
Practice 4: $x = 0$
1. $x = 6$
4. $x = 3$
8. $x = 41$
12. $x = 16$
16. $x = -1$
20. $x = 4$
24. $x = 6.3$
28. $x = -48$

Unit 60

Practice 1: $x = -8$
Practice 2: $x = -18$
1. $x = 4$
4. $x = 8$
8. $x = -4$
12. $x = 9$
16. $x = 18$
20. $x = 36$
24. $x = 32$
28. $x = -42$

Unit 61

Practice 1: $x - 8y$; $12x$
Practice 2: 42
Practice 3: $-3x - 12y$
1. $7x$
4. $10a$
8. $8a + 2$
12. $3y$
16. 36
20. –6
24. $-4x - 4y + 4z$
28. $4p - 80$

Unit 62

Practice 1: $x = -7$
Practice 2: $x = 0$
1. $x = 8$
4. $x = 6$
8. $x = \frac{1}{4}$
12. $x = 10$
16. $x = 1$
20. $x = -2$
24. $x = \frac{3}{2}$
28. $x = -\frac{2}{5}$

Unit 63

Practice 1: $75
Practice 2: Rene: 4 children; Diane: 8 children
Practice 3: 20 sets dishes; 60 sets pots

1. $125.00
3. Micro = 25 computers; Morse = 5 computers
5. 100 meatball; 200 cheese
7. 20 sleepers; 10 nonsleepers
9. $3,430.00
11. Nanda = 90 homes; Lane = 30 homes
13. $214.00
15. Bill's = 450 sandwiches; Pete's = 90 sandwiches
17. 240 diskettes; 48 boxes of paper
19. 420 thermometers; 60 hot water bottles

Chapter 5 Test

1.

2a. negative
2b. negative
2c. positive
2d. positive
3a. 12
3b. –16
3c. –27.6
3d. $\frac{2}{3}$
4. –27
5. –11
6. 19
7. 105
8. $-\frac{4}{3}$
9. 14
10. $5x - 3$
11. 9
12. $3x$
13. $-4x + 28$
14a. $x = 28$
14b. $x = 9$
14c. $x = -26$
14d. $x = 3$
15. Pups = 3 runs
Seals = 9 runs

APPENDIX

Chapter Organizers

CHAPTER ORGANIZER 1

Whole Numbers

Key Topics	Key Points	Examples
Writing and reading whole numbers.	The position of the digits determines place values. Compound numbers are 21 through 99 and are hyphenated when written as words.	1. The number 283 is "two hundred eighty-three." 2. Four thousand seven is "4,007."
Rounding whole numbers.	The process of determining approximate values.	The number 6,718 rounded to the nearest thousand is 7,000.
Adding and subtracting whole numbers.	The process of finding sums and differences.	1. 9,604 + 127 = 9,731 2. 4,392 − 257 = 4,135
Multiplying whole numbers.	The process of repeated addition of the same number. Answers are called *products*. Numbers being multiplied are called *factors*.	316 × 42 632 1,264 13,272
Dividing whole numbers.	The process of determining how many times a number is contained in another as a multiple. Division by 0 is impossible. Division into 0 always equals 0.	16)1,402 = 87 128 122 112 10 Answer: 87 R10

CHAPTER ORGANIZER

2 Fractions

Key Topics	Key Points	Examples
Fraction types and fraction parts.	Fractions are numbers representing parts of a whole and are made up of a numerator, denominator, and sometimes a whole number. Types of fractions are *proper, improper,* and *mixed numbers.*	1. $\frac{3}{4}$ ← Numerator, ← Denominator 2. $2\frac{3}{4}$ is a mixed number. 3. $\frac{3}{4}$ is a proper fraction. 4. $\frac{4}{3}$ is an improper fraction.
Reducing fractions.	The process of changing the form of a fraction without changing its value.	$\frac{36}{52} = \frac{9}{13}$
Adding and subtracting fractions.	Only fractions with like denominators can be added or subtracted. The lowest common denominator is needed when combining fractions with unlike denominators.	1. $4\frac{1}{2} = 4\frac{4}{8}$; $+\ 2\frac{3}{8} = 2\frac{3}{8}$; $= 6\frac{7}{8}$ 2. $5\frac{4}{7} = 5\frac{8}{14}$; $-\ 3\frac{1}{2} = 3\frac{7}{14}$; $= 2\frac{1}{14}$
Multiplying fractions.	Common denominators are unnecessary. Multiply numerators by numerators and denominators by denominators. Rewrite mixed numbers as improper fractions and whole numbers over 1 before multiplying. Use cancellation whenever possible.	$4 \times \frac{2}{3} \times 5\frac{1}{4} \times 2$ $= \frac{\cancel{4}^{1}}{1} \times \frac{2}{\cancel{3}_{1}} \times \frac{\cancel{21}^{7}}{\cancel{4}_{1}} \times \frac{2}{1}$ $= \frac{28}{1} = 28$
Dividing fractions.	The process of multiplying by the reciprocal of the divisor.	$6\frac{2}{3} \div 1\frac{1}{3}$ $= \frac{\cancel{20}^{5}}{\cancel{3}_{1}} \div \frac{\cancel{4}^{1}}{\cancel{3}_{1}} = \frac{20}{3} \times \frac{3}{4}$ $= \frac{5}{1} = 5$

CHAPTER ORGANIZER
3
Decimals

Key Topics	Key Points	Examples
Reading and writing decimals.	The position of the digits determines place values. The decimal point is read as "and." Digits to its left express the whole number part. Digits to its right express the fractional part.	1. The number 1.016 is written as "one and sixteen thousandths." 2. Three and forty-one hundredths is read as "3.41."
Rounding decimals.	The process of approximating decimal values.	The number 4.37 rounded to the nearest tenth is 4.4.
Adding and subtracting decimals.	Line up decimal points and numbers of the same place value. Place the decimal point in the sum or difference directly below the decimal points in the column.	6.032 + .0072 is $\begin{array}{r} 6.032 \\ +\ \ .0072 \\ \hline 6.0392 \end{array}$
Multiplying decimals.	Arrange and multiply decimals as if they were whole numbers. Place the decimal point in the product by counting from the right the number of decimal places to the right of the decimal points in the multiplier and the multiplicand. Add zeroes as necessary.	$\begin{array}{r} 4.17 \\ \times\ 3.1 \\ \hline 417 \\ 1{,}251\ \ \\ \hline 12.927 \end{array}$
Dividing decimals.	Move the decimal point in the divisor to the right of the last digit when necessary. Move the decimal point in the dividend an equal number of places and bring it directly above into the quotient. Divide as in whole numbers. Add zeroes and round the quotient as necessary.	$3.2\overline{)14.72} = 32\overline{)147.2}$ with quotient 4.6; $\begin{array}{r} 128 \\ \hline 192 \\ 192 \\ \hline 0 \end{array}$
Converting decimals to fractions.	Accurately read the decimal out loud and write what is heard.	$4.003 = 4\frac{3}{1{,}000}$
Converting fractions to decimals.	Divide numerators by denominators. Add zeros and round whenever necessary.	$\frac{7}{8} = 8\overline{)7.000}$ with quotient .875

CHAPTER ORGANIZER 4

Percents

Key Topics	Key Points	Examples
Converting decimals and fractions to percents.	Change any fractions to decimals. Move the decimal point two places to the right and add a percent symbol.	1. $\frac{3}{4} = .75 = 75\%$ 2. $2.3 = 230\%$
Converting percents to decimals.	Express any fractional part of the percent as a decimal. Drop the percent symbol and move the decimal point two places to the left.	$4\frac{1}{4}\% = 4.25\% = .0425$
Converting percents to fractions.	Express any decimal part of the percent as a fraction and change any mixed numbers to improper fractions. Divide by 100 (multiply by $\frac{1}{100}$) and reduce to lowest terms.	$6\frac{1}{4}\% = \frac{25}{4} = \frac{25}{4} \times \frac{1}{100} = \frac{25}{400} = \frac{1}{16}$
Percent formulas.	1. *Base* is the beginning quantity. 2. *Portion* is the quantity compared with the base. 3. *Rate* is the percent. $\text{Base} = \frac{\text{Portion}}{\text{Rate}}$ $\text{Portion} = \text{Rate} \times \text{Base}$ $\text{Rate} = \frac{\text{Portion}}{\text{Base}}$	1. 12 is 50% of what number? $\text{Base} = \frac{12}{50\%} = \frac{12}{.50} = 24$ 2. 6% of 12 is what number? $\text{Portion} = 6\% \times 12 = .06 \times 12 = .72$ 3. 42 is what percent of 84? $\text{Rate} = \frac{42}{84} = .5 = 50\%$

CHAPTER ORGANIZER 5

Introduction to Algebra

Key Topics	Key Points	Examples
Signed numbers.	Positive numbers are greater than zero. Negative numbers are less than zero.	Graph –2 and 5 on a number line. –5 –4 –3 –2 –1 0 1 2 3 4 5
Adding and subtracting signed numbers.	To add signed numbers of the same signs, add the numbers and keep their common sign. To add signed numbers of different signs, disregard the signs and subtract the smaller number from the larger. Attach the larger number's sign to the answer. To subtract signed numbers, change the sign of the number to the right of the subtraction symbol to its opposite. Change the subtraction sign to addition and apply the rules for addition of signed numbers.	1. $-6 + (-3) = -9$ 2. $-4 + 6 = 2$ 3. $6 - (-7) = 6 + (7) = 13$ 4. $-6 - (-7) = -6 + (7) = 1$
Multiplying and dividing signed numbers.	$P \times P = P$, $P \times N = N$, $N \times P = N$, $N \times N = P$ and $P \div P = P$, $P \div N = N$, $N \div P = N$, $N \div N = P$	1. $(-6) \times (-3) = 18$ 2. $(-12) \div (-4) = 3$ 3. $6 \times (-3) = -18$ 4. $12 \div (-4) = -3$
Distributive property.	Distribution is a way of expanding expressions using multiplication. Multiply each term inside the set of parentheses by the multiplier outside the set of parentheses.	$3(x + y) = 3x + 3y$
Order of operations.	1. Do all operations within grouping symbols. 2. Do all operations involving exponents or roots. 3. Do all multiplication or division from left to right. 4. Do all addition or subtraction from left to right.	$(12 - 3) + 4 \div 2 + 3^2$ $= (9) + 4 \div 2 + 3^2$ $= 9 + 4 \div 2 + 9$ $= 9 + 2 + 9$ $= 20$
Similar terms.	Only similar terms can be combined. Similar terms are those whose variables are exactly alike.	$4x + 3y + (-7x) = -3x + 3y$

(Concluded on next page.)

CHAPTER ORGANIZER

5 Introduction to Algebra

Key Topics	Key Points	Examples
Solving equations.	1. Remove parentheses by distribution. 2. Combine similar terms. 3. Isolate algebraic terms on one side, constants on the other side. 4. Divide or multiply to solve the equation. The same number can be added, subtracted, multiplied, or divided on both sides of the equation without affecting its equality.	Solve for x. $2(x + 2) = 8 - 2x$ $= 2x + 4 = 8 - 2x$ $2x + 4 = 8 - 2x$ $-4 \quad -4$ $2x = 4 - 2x$ $+2x \quad +2x$ $4x = 4$ $\frac{4x}{4} = \frac{4}{4}$ $x = 1$

Index

Notes

Notes

Notes

Notes